"ASPECT" GEOG[RAPHIES]

TROPICAL GEO[GRAPHY]

Other titles in the "Aspect" Series

A GEOGRAPHY OF MANUFACTURING
A GEOGRAPHY OF SETTLEMENTS
BIOGEOGRAPHY
A GEOGRAPHY OF TOURISM

"ASPECT" GEOGRAPHIES

TROPICAL GEOGRAPHY

An Introductory Study of the Humid Tropics

H. R. JARRETT, B.A., M.Sc.(Econ.), Ph.D., F.R.G.S.

*Formerly Senior Lecturer in Geography at
Fourah Bay College, Freetown, and at University College, Ibadan
(now the University of Ibadan)*

MACDONALD AND EVANS

MACDONALD & EVANS LTD
Estover, Plymouth PL6 7PZ

First published 1977

©
MACDONALD AND EVANS LIMITED
1977

ISBN: 7121 2018 1

This book is copyright and may not be
reproduced in whole *or in part* (except for
purposes of review) without the express
permission of the publishers in writing.

Filmset in Photon Times 11 on 12 pt by
Richard Clay (The Chaucer Press), Ltd, Bungay, Suffolk
and printed in Great Britain by
Fletcher & Son Ltd, Norwich

Introduction to the Series

THE study of modern geography grew out of the medieval cosmography, a random collection of knowledge which included astronomy, astrology, geometry, political history, earthlore, etc. As a result of the scientific discoveries and developments of the seventeenth and eighteenth centuries many of the component parts of the old cosmography hived off and grew into distinctive disciplines in their own right as, for example, physiography, geology, geodesy and anthropology. The residual matter which was left behind formed the geography of the eighteenth and nineteenth centuries, a study which, apart from its mathematical side, was encyclopaedic in character and which was purely factual and descriptive.

Darwinian ideas stimulated a more scientific approach to learning, and geography, along with other subjects, was influenced by the new modes of thought. These had an increasing impact on geography, which during the present century has increasingly sought for causes and effects and has become more analytical. In its modern development geography has had to turn to many of its former offshoots—by now robust disciplines in themselves—and borrow from them: geography does not attempt to usurp their functions, but it does use their material to illuminate itself. Largely for this reason geography is a wide-ranging discipline with mathematical, physical, human and historical aspects: this width is at once a source of strength and weakness, but it does make geography a fascinating study and it helps to justify Sir Halford Mackinder's contention that geography is at once an art, a science and a philosophy.

Naturally the modern geographer, with increasing knowledge at his disposal and a more mature outlook, has had to specialise, and these days the academic geographer tends to be, for example, a geomorphologist or climatologist or economic geographer or urban geographer. This is an inevitable development since no one person could possibly master the vast wealth of material or ideas encompassed in modern geography.

This modern specialisation has tended to emphasise the importance of systematic geography at the expense of regional geography, although it should be recognised that each approach to geography is incomplete without the other. The general trend, both in the universities and in the school examinations, is towards systematic studies.

This series has been designed to meet some of the needs of students pursuing systematic studies. The main aim has been to provide introductory texts which can be used by sixth-formers and first-year university students. The intention has been to produce readable books which will provide sound introductions to various aspects of geography, books which will introduce the students to new ideas and concepts as well as more detailed factual information. While one must employ precise scientific terms, the writers have eschewed jargon for jargon's sake; moreover, they have aimed at lucid exposition. While, these days, there is no shortage of specialised books on most branches of geographical knowledge, there is, we believe, room for texts of a more introductory nature.

The aim of the series is to include studies of many aspects of geography embracing the geography of agriculture, the geography of manufacturing industry, biogeography, land use and reclamation, food and population, the geography of settlement and historical geography. Other new titles will be added from time to time as seems desirable.

H. ROBINSON
Geographical Editor

Author's Preface

THIS book is intended to be an introductory study of the geography of the humid tropics, a term which is now in fairly general use and which is explained in the chapters which follow. As the regional geography which was so characteristic of a former age is inexorably disappearing from examination syllabuses (for better or worse) its place is being taken by "systematic" studies—studies in various branches of the subject. The physical side of geography has in fact long been systematised within the general subjects of Geomorphology and Climatology but the human aspects were much slower to assert their claim to similar treatment. This was probably inevitable as geographers, for the most part unwittingly, awaited the growth of the social sciences in their own right. The very considerable developments which have taken place over the past few decades in these sciences have in recent years had a tremendous impact upon geographical thought, and this is perhaps particularly the case with regard to Economics, a point which will become very clear to the reader of this book. It is hoped that while the following pages present some of the complications of the geography of the humid tropics to the reader they will not confound him with complexity and that they will afford him at least a few glimpses into a fascinating and developing subject.

November 1976   H.R.J.

Introduction

IT might conceivably be argued with some force that while yesterday and perhaps even today belong to the lands of the cool temperate zone, tomorrow will belong to the humid tropics. This is not in itself a farfetched notion, for after all, the foci of civilisation and power have shifted latitudinally over the earth's surface in times past and there seems no good reason why they should not do so again. There is certainly no doubt that the status and the development of tropical areas are today topics which command widespread interest, and there are several reasons for this.

In the first place, we now possess a very considerable amount of information about most tropical territories and about tropical regions generally; viewers can see on their television screens contemporaneous events taking place in these regions, and people generally are more knowledgeable about the tropics than they have ever been in the past. It was in fact as long ago as 1848, more than a century and a quarter ago, that *The Times* proclaimed that "a man needs to rub up his geography these days." Today the observer of current events has it brushed up for him—frequently in colour—on the small screen in the privacy of his own home, and it is almost inevitable that this increasing knowledge begets increasing interest in the minds of those mentally alert as well as in more specialised circles as research workers press on seeking to make further scientific advances. Where there is little knowledge, and when opportunities of furthering it are very limited, interest is perforce at least partially stifled—possibly entirely so—but that is no longer the situation with regard to the humid tropics, for our knowledge of this part of the earth's surface has tremendously increased in recent years and is still rapidly growing.

From increasing knowledge it is but a step to increasing concern. We can no longer lightly dismiss from our minds, for instance, the fact that millions of this world's inhabitants (very many of whom live in the humid tropics) live in poverty and in a state of almost continuous fear of hunger; it is not surprising that more and more responsible folk are actively concerned to discover means by which poverty and hunger can be alleviated. This can sometimes be achieved by the application of improved agricultural techniques and sometimes by the judicious establishment of industry, both of which topics are dealt with in this book. And supporting all other forms of progress there must be an extending educational system. All these topics command considerable interest in our own community life, and it is not therefore surprising that a corresponding interest in these and other problems of development which today face the tropical world is now strongly developed, for these problems possess a powerful intrinsic interest to an ever-increasing number of observers.

The purpose of this book is to bring to the attention of the reader some of the important facts about the humid tropics, and also, more importantly, some of the problems which the inhabitants of these regions are today grappling with. It is an introductory work especially prepared for readers coming to the serious study of the subject for the first time. On the other hand, the problems examined are looked at in some detail and not in a merely superficial manner. Many readers, possibly most of them, will be students preparing for various examinations, and it is hoped that the material here presented will prove useful in the preparation for examinations, particularly those in geography which have at least in part a bias towards the geography of tropical lands. Although the writer has had considerable first-hand experience of the humid tropics, it is inevitable that a work of this nature draws freely upon many authorities, and acknowledgment is readily made, especially to those writers whose works appear in the bibliography. This feature of the book may also be of service in directing the interested student to more detailed and specialist works.

No apology is made for the fact that the treatment offered in this volume is topical and not regional. The geography texts of an earlier age were frequently regional in outlook, treating individual territories separately and thus, at least implicitly, emphasising the differences which inevitably exist between them. While this outlook persisted it was virtually impossible to build up any coherent body of theory, except, interestingly enough, in climatology and physical geography, both subjects which flourished in their own right. The treatment of climate has always been largely topical, and while the physical geographies of various territories were treated regionally they did draw heavily upon the growing discipline of geomorphology. The relative strengths of climatology and geomorphology in fact reflected in large measure the interests of many of the founding fathers of geography who came to the subject from other scientific areas of study such as geology.

The topical treatment has the advantage that it emphasises the likenesses as well as the differences between various territories, and it becomes less deterministic as material from the social sciences is more readily used, and this brings a greater measure of realism into what, after all, is essentially a "human" subject. It is hardly too much to say that the role of physical geography *within the parent subject* (not in its own right as a scientific discipline) is today a noticeably diminishing one, and this is likely to continue as geographers increasingly realise how much subjects like economics can contribute to their patterns of thinking.

The adoption of this topical viewpoint does not imply any belief that regional geography can now be dispensed with as a form of study. On the contrary, it remains of great importance if our subject is not to become overwhelmingly theoretical, divorced from that very ground upon which it is supposed firmly to stand. Numerous regional examples are examined in the following pages, but the wise student will make good use of some of the excellent regional geographies of different parts of the humid tropics which are available to complement his study of this present book.

A perusal of the Table of Contents will quickly reveal the general plan of the book. After a survey of the location and the general nature of the humid tropics comes a study of the human situation, including the important topic of population. Thus the reader gains at the outset an overall picture of the dominant physical and human characteristics of the tropics. Because most of the inhabitants of the region are still farmers a study of farming systems follows, and attention is paid to the urgent problem of improving upon or even replacing traditional farming practices. Since it is frequently assumed that industrialisation brings with it increasing prosperity it is natural that next in order we should examine the possibilities for industrial development.

Because the region dealt with is a very extensive one, it is inevitable that many variations, both within the physical and the human environments, occur within it. This is fortunate, as it means that by the sort of economic co-operation which we call trade individual territories can work together for the good of all, while trading relationships have also been established between tropical territories on the one hand and territories of the temperate latitudes on the other. It is more than three centuries since John Donne reminded us that "No man is an Iland, intire of its selfe," and this is if anything more true today than it has ever been; it is therefore right that we conclude this study with an examination of some aspects of man's attempts within the humid tropics to improve his standards of living through transport, markets and trade. It is, after all, very largely through trade that civilisations and prosperity come, for the community which is self-contained is severely limited in the range of its ideas and in its material well-being.

The reader will observe that mention is frequently made in these pages of the L.D.C.s—the "less developed countries" as they are now called. While it is true that many L.D.C.s lie outside the humid tropics, the greater number lie within that great region, and since L.D.C.s everywhere have much in common their problems are very largely the problems of the tropical world. If this work helps in any way to increase knowledge of and interest in what is a fascinating part of the world it will have achieved its purpose.

Contents

Chapter		Page
	Introduction to the Series	v
	Author's Preface	vii
	Introduction	ix
	List of Illustrations	xv
I.	The Region Defined	1
	The human background	1
	The physical background	11
	Further climatic considerations	27
II.	Natural Resources of the Humid Tropics	31
	The role of the environment	31
	The present situation	33
	Resources and the environment	34
	The assessment of natural resources	37
	Water	39
	Soils	42
	Vegetation	49
	Minerals and power	52
III.	Human Resources of the Humid Tropics	58
	Preliminary considerations	58
	The past and the present	58
	Climate and the human factor	62
	Disease	65
	Diet and nutrition	70
	Population	74
	Population control	84
IV.	Agriculture in the Humid Tropics: Shifting Cultivation	87
	Shifting cultivation	87

Chapter		Page
	A case study: the Maya	100
	Final considerations	103
V.	AGRICULTURE IN THE HUMID TROPICS: OTHER FORMS OF FARMING ACTIVITIES	105
	Wet-land cultivation	105
	Cash crop farming	107
	Plantations	113
	Livestock farming	128
VI.	INCREASING AGRICULTURAL PRODUCTIVITY IN THE HUMID TROPICS	133
	The problem examined	133
	Agricultural productivity in the humid tropics	139
	Land	141
	Labour	146
	Capital	147
	Enterprise	157
VII.	THE BACKGROUND TO INDUSTRIALISATION IN THE HUMID TROPICS	159
	The early stages of industrial development	159
	The present situation	163
	Agriculture and industry	165
	The economic bases of industrialisation	168
	Capital and productivity	172
	The balance of development	173
	Some other problems of industrial development	176
	The roles of agriculture and industry	178
	What is the answer?	180
	The pattern of industry	182
VIII.	TRANSPORT, MARKETS AND TRADE IN THE HUMID TROPICS	184
	Transport	184
	Markets	197
	Trade	201
	Final comments	210
	BIBLIOGRAPHY	213
	INDEX	217

List of Illustrations

Fig.		Page
1.	*Per capita* incomes in the humid tropics (1970)	5
2.	The humid tropics	13
3.	Climatic graphs for Manaos, Bombay, Freetown and Kayes	14
4.	The tropical humid climates: I	15
5.	The tropical humid climates: II	17
6.	The relation between plant communities and relief in the wet-and-dry tropics: I	24
7.	The relation between plant communities and relief in the wet-and-dry tropics: II	25
8.	A typical rural scene in the Deccan	42
9.	Relief and soils in the humid tropics	45
10.	A possible pattern of malaria control	69
11.	The world: natural population increase	76
12.	Demographic transition in England and Wales (1700–1970)	77
13.	Age and sex pyramids of three tropical territories	80
14.	The low-level equilibrium population trap	84
15.	Farming calendars in the humid tropics	91
16.	West Africa: types of subsistence economy	92
17.	The Mayan Empires (after Gourou)	101
18.	Cocoa prices and political changes in Ghana (1950–1966)	112
19.	Banana production in the Caribbean	123
20.	Effects of nitrogenous fertiliser upon cereal yields in India	150
21.	The development continuum	166
22.	Capital formation and income in developing countries	169
23.	An example of transport in Africa	187
24.	The proposed Trans-African Highway	196
25.	The theory of periodic markets	199

Chapter I

The Region Defined

THE tropical world is vast. Many observers will readily accept that general statement, but those whose notions of global areas are based upon Mercator's famous projection are bound to receive it with some surprise, for this projection, excellent as it is for showing direction, is grossly misleading with respect to areas. It wildly exaggerates the areas of extra-tropical regions and correspondingly it represents the tropical parts of the world as being far smaller than they actually are. In fact, about 42 per cent of the total surface area of the globe lies between the latitudes of 25 degrees north and 25 degrees south, and considerably more than half of this total can be regarded as falling within the humid tropics. Not all of this vast region, of course, is land; rather more than three-quarters is occupied by oceans and seas.

THE HUMAN BACKGROUND

Many, perhaps most, of the inhabitants of the "developed" world have more or less stereotyped ideas about their counterparts in the humid tropics. They wear, for example, a minimum of clothing and live in thatched mud huts; they are very poor and possess few, if any, of the amenities of life which we in the industrialised countries are apt to take for granted; and they are frequently hungry, if not actually starving. And while, like all generalisations, these particular ones are open to criticism, it must be admitted that they do contain a substantial element of truth. While direct comparisons between living standards and conditions of life in the humid tropics and those in temperate lands are usually very misleading, it is true that for most of the inhabitants of the tropical world life is burdensome beyond anything we ourselves know. Grigg (1970, 24),* for instance, quotes an FAO estimate that about one-tenth of the world's population is "undernourished" and about one-half "malnourished," and it seems that most of these unfortunate

---

*The reader should refer to the Bibliography for publication details of the works cited.

souls live in the tropics, while many readers will be familiar with the famous statement made by Lord Boyd Orr (in 1950 when he was Sir John Boyd Orr) that "a lifetime of malnutrition and actual hunger is the lot of at least two-thirds of mankind." The accuracy of statements such as these have more recently been sharply questioned and we shall refer to this in Chapters III and VI, but there can be no doubt whatever that large numbers of the world's people do not have enough to eat, and unfortunately dietary deficiencies such as these are widespread in the humid tropics where they lead, in their turn, to apathy and diminished food production (and therefore to an even greater degree of malnutrition), together with a heightened susceptibility to disease. So a kind of vicious circle is created, and it is very difficult for most of the people concerned to escape from this.

Such deficiencies are emphasised by the general lack of many of the amenities of life which most of the inhabitants of the "Western" world would consider essential. Miller (1952, 21), for instance, points out that in northern Nigeria most of the inhabitants lack adequate housing, clothing and blankets, and they therefore suffer greatly from the numbing physiological cold produced by temperatures which on December and January nights often fall below 4·4° C (44° F), coupled with relative humidities which are frequently below 30 per cent. Conversely, the body heat-regulating mechanism is under heavy strain when temperatures exceed that of the body, and this normally is the case during the hot season, particularly in tropical continental regions which have a lengthy dry season.

It is in the ordinary day-to-day living that very great disparities are apparent between the quality of living experienced by most people in the tropics and that enjoyed by those whose home is in the "developed" territories. In the tropics houses or huts usually are fitted with few, if any, of the modern appliances (cupboards, sinks, baths, etc.) with which we are so familiar; "specialist" rooms (kitchen, dining room, bedrooms, etc.) are hardly known, while cooking is performed over open fires—probably out of doors whenever possible. Living conditions around the peripheries of cities can be crowded and insalubrious. Clarke (1974a) has reminded us, for instance, of urban living conditions in the Caribbean. Dwellings frequently consist of single-roomed wooden sheds, often perched precariously on hillsides, while there are no proper means of sewage disposal. Covered sewers are often lacking and insanitary pit-latrines are in common use in poorer quarters. So many inhabitants live under these desperate conditions that in Kingston, Jamaica, town planners do not consider that "overcrowding" exists until densities exceed two persons per room or eight persons per water closet. Squatting is widespread on any site outside the urban land market area, on abandoned lots, in ravines, on swampy ground and on near-precipitous slopes.

Unfortunately this kind of situation is all too common, and the depressing story could be extended almost indefinitely—to Brazil, for example, where São Paulo has been described (*The Times*, 15th August 1975) as an overcrowded concrete jungle with 4·5 m^2 of living space per inhabitant, and with an atmosphere which is so polluted that everyone who lives there has a smoker's cough whether he smokes or not. It seems that there is often little to choose in terms of comfort between city life in the tropics and life in the country where water may have to be fetched and carried in kerosene tins for varying distances— in some instances as far as 2 km at the height of the dry season. This is normally women's work. Men's work is also laborious. Instead of the glossy office or even the factory floor, both reached in the comparative comfort of the owner-driven car, there is the insect-ridden bush to be cleared by hand labour after a lengthy and wearisome walk under the hot sun—hard work indeed for those who are not only undernourished but also quite possibly further debilitated by disease.

It can be salutary and even shaming to compare this distressing lot of many of the inhabitants of the tropical world with that which obtains in the wealthier territories. Ward and Dubos (1972, 174–5) point out, for example, that every American baby born will statistically consume, before his death, over four million joules a year in various kinds of foods as well as 12,250 litres of petrol. To sustain the overall incredible consumption rate within the U.S.A., which numbers about 5 per cent of the total world population, no less than 27 per cent of the world's total resource production is needed.

It is very easy at this point for Western minds to be afflicted by a form of guilt, but it is questionable whether this is entirely justified. It is comparatively easy to overlook the other parts of the world picture. It is not true, for example, that all the people of the humid tropics live under the depressing conditions so far described—many of them, in fact, live under conditions of considerable luxury—while it is on the other hand true that for many inhabitants of the developed states life is almost unbearably hazardous and burdensome. And while we can point, as we have done, to rates of consumption in the developed world which appear excessive we should remember that the developed group of territories together account for more than two-thirds of the total world product, and that this considerable feat has been achieved through hard work and foresight.

The lesson should be clear. There is no future in pity, admirable as this emotion is in its place, nor in charity. What is desperately needed is skilled help which will enable the L.D.C.s to increase their productivity. Bairoch believes that the levels of agricultural productivity in the L.D.C.s are 45 per cent below those attained by the now developed countries at the start of the Industrial Revolution, and that the implied gap is being bridged by imports of foodstuffs from the developed parts of the world. This is a tremendously dangerous situation when the developed countries are finding it harder to feed themselves and when the cost of food imports is therefore rising sharply. An observer writing in *The Statesman*, published in Calcutta, has warned his fellow-countrymen: "if you are poor you have only yourself to blame. Development is a matter of hard work and discipline." While we may not agree entirely with this, there is much truth in it, though we must recognise that unfortunately hard work and discipline can yield disappointing results if they are not wisely and productively applied.

We should also bear in mind that the depressing picture which we have so far painted of life in the humid tropics is constantly being modified, perhaps more rapidly than is sometimes thought. More and more work is being created in offices, factories, shops and elsewhere for all classes of salaried workers. But the dominant factor is that of finance; the rate of development from a traditional society towards a developed one is bound to be slow until the economy of the country concerned is strong enough to sustain the change. This important point will receive considerable attention in later pages.

Definitions of Underdevelopment

It is desirable at this point to define more clearly what we mean by a less developed country, and we must acknowledge right away that there is no simple method of doing this. There are, however, certain characteristics which appear to be fairly general throughout the tropical world, and which can be said together to constitute "underdevelopment."

1. *Per capita* incomes are low by comparison with those of the "developed" nations. Fig. 1 gives some idea of the present situation, and it shows all territories which have substantial areas within intertropical latitudes. Not all of these territories, of course, fall within the compass of this book. It can hardly be argued, for instance, that Australia and the Republic of South Africa are tropical countries, while the vast Republic of Sudan does not fall within the humid

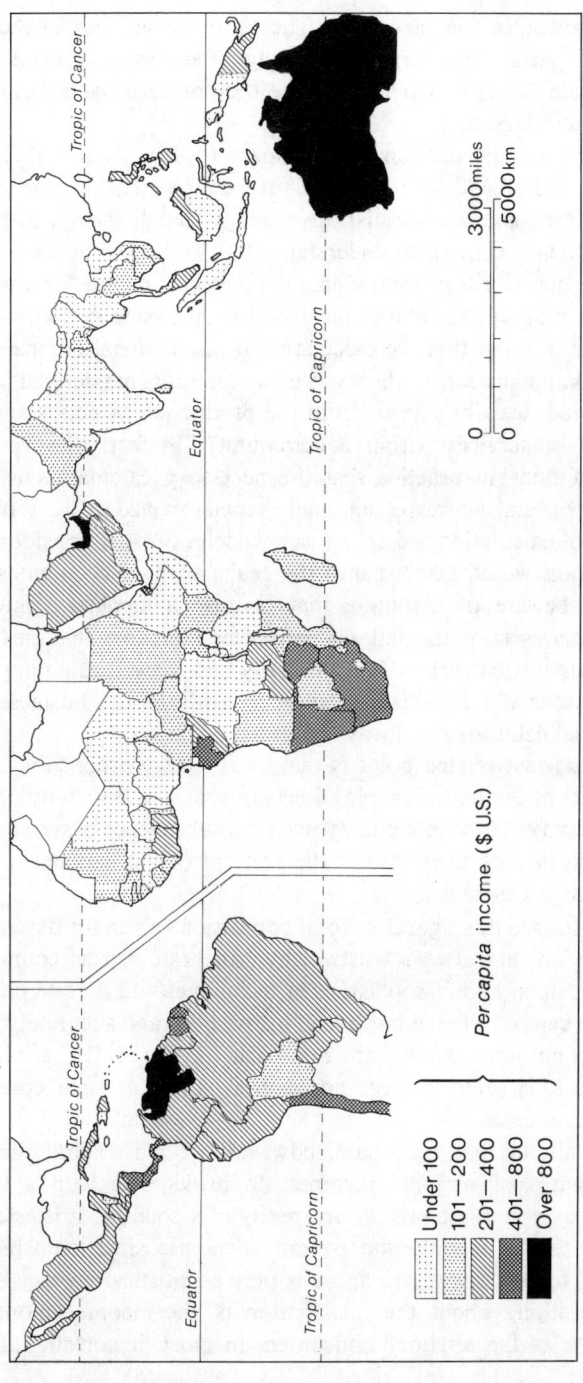

FIG. 1.—Per capita incomes in the humid tropics (1970).

tropics except in the far south. The more heavily populated and developed parts of the territory lie roughly across the central zone, well outside the humid tropics, while the northern parts form part of the Sahara Desert.

Some words of caution with respect to Fig. 1 are obviously desirable. We should bear in mind, for example, that the concept of *per capita* income is a sophisticated one in detail, though in broad terms it is easy enough to understand. It is arrived at in the case of any particular territory by dividing the national income for a period such as a year by the total population for that same period, but we must bear in mind that the calculation of the national income itself is not a simple matter. In theory it is the total income derived over a given period, usually a year, from the production of goods and the provision of services within a community. In fact, however, the economist finds in practice that the necessary calculation is very complicated, and figures of national income arrived at *via* different methods of calculation are apt to vary widely. Further consideration of this point would take us into the realm of pure economics, but we must beware of assuming that because a monetary figure is given as representing the national income of a territory that the figure is completely trustworthy. It is probably best viewed as an approximation, especially in L.D.C.s where records in the business and commercial fields are not always accurate.

An important related point is that a very large proportion of the inhabitants of the humid tropics does not share, at least fully, in any cash economy. These inhabitants live on a subsistence basis, and any attempt to include them within the concept of the national income must be largely artificial.

It is also true that figures of total populations given for developing territories are not always trustworthy. Accurate census counts are dependent upon highly sophisticated techniques and literate populations who can be relied upon to read instructions and accurately fill in forms—conditions which do not exist in the L.D.C.s. Again, therefore, it is safer to treat population figures of such countries as approximations.

When all this has been said, however, we must admit that *per capita* figures of national incomes do provide us with a useful means of assessing the *overall* prosperity of a country. It is essential to stress that it is only the overall economic situation which is indicated, for the *per capita* figure is only a statistical average and it tells us nothing about the distribution of the income among the inhabitants of the territory concerned. In most, if not all, L.D.C.s there are very great differences between the highest and the lowest incomes, and a misleading impression of personal prosperity can be given by a high *per capita* national income figure. The figure for Venezuela, for instance, is now over $1000 *per annum*, which is a high figure for a developing country, but tremendous extremes exist between the affluent minority and the majority who still live under conditions of extreme poverty. The "spaghetti junctions" and the skyscrapers of the capital, Caracas, stand in gaunt contrast to the shanty towns of the urban periphery, while agriculture remains backward and industry generally is inefficient.

While the situation in Venezuela is admittedly extreme because of its tremendous oil wealth, it represents a situation which is widespread in the humid tropics, and the resulting total of human distress is great indeed. In parts of southern India, for instance, a young baby can even today be sold by the starving father to adoptive parents for less than ten rupees—roughly half the price of a chicken! Conditions of chronic mass poverty are still widespread in the developing world, and it is mainly in pockets of economic development (*see* 2 below) that this fearful situation is alleviated.

2. The potential for future economic expansion in the L.D.C.s is normally very considerable because resources are not at present used to anything like their fullest extent. In other words, the prevailing poverty to which we referred in 1 above is not essentially due to excessively limited natural resources. It is also typical of L.D.C.s that their development is ill-balanced, and such countries frequently rely on a small number of cash products to support the economic and social life of the community. This can be clearly demonstrated by a consideration of exports; the following table shows examples drawn from selected African territories which rely heavily upon a single export or a small number of exports.

Country	Product(s)	% by value of all exports
Gambia	Groundnuts and groundnut products	95
Zambia	Copper	95
Chad	Cotton	83
Senegal	Groundnuts and groundnut products	78
Liberia	Iron ore	75
Ghana	Cocoa	69

It is also typical of territories at an immature stage of development that they exhibit "pockets" of comparatively advanced economic development (called "islands of development" by Hance—1964, 46–50) set in a matrix of traditional production. Further development comes as these pockets, which may be agricultural, mining or industrial in nature, expand, or as new ones come into being. Production then becomes correspondingly less ill-balanced and more mature.

3. L.D.C.s are also characterised by the continuing use of obsolete and traditional methods of production and by outmoded forms of social organisation. It is significant that most Latin American countries discovered in practice that an essential pre-requisite to economic development was the overthrow of established historically-based dictatorships as a preliminary to the redistribution of land, to the introduction of modern methods of agriculture, and to the establishment of industry. Another case in point is that of Ethiopia, a country which has remained extremely backward and feudal, which only in 1975 overthrew her emperor, though at the time of writing it is too early to say just what form the new pattern of government will take.

4. The occupational structure of L.D.C.s is characterised by its immaturity. In a traditional society by far the greater number of workers and their families are dependent for their livelihood upon primary production, mainly upon agriculture. In extreme cases over 90 per cent of the working population are engaged upon the land. It has been suggested that one of the hall-marks of an L.D.C. is that it has more than 50 per cent of its workers engaged in agriculture, forestry and/or fishing. The corollary is that in such territories the secondary and tertiary sectors of the economy are in the earliest stages of development.

The above are all points which will receive more detailed treatment in later pages, but we should bear in mind that they are by no means the only criteria by which the developmental status of a territory can be assessed. Other possibilities include small amounts of energy consumed per head of population, a low *per capita* use of water, and a low literacy rate. But one fact is clear: the gap between the richest and the poorest countries is desperately wide, the *per capita* income in the poorest territories being less than one-twentieth of that of many developed states.

Hodder (1973) suggests that the most satisfactory way of defining underdevelopment may be empirical. This is possible if we can isolate certain characteristics which are typical of L.D.C.s in general, though they may not all occur in every single case. If we adopt this line of approach we might recognise the following main features which complement and emphasise those previously mentioned.

1. Low levels of nutrition and hygiene are normal and this results in generally poor health, a high infant mortality rate and a low life expectancy.

2. The traditional methods of production which are dominant are unfortunately inefficient and static; they do not easily lend themselves to the modifications essential for the emergence of a more efficient and dynamic economy. Communal ownership of land is widespread and personal attitudes are frequently conservative. There is very little diversification of economic activity, and agricultural pursuits are dominant.

3. The continuing employment of outmoded forms of production brings in its train chronic and widespread poverty.

4. Because of the generally low level of incomes there is a low level of market demand. This is not helpful to the establishment of a cash economy, a trend which is strengthened by the high proportion of productive workers engaged in the primary fields of agriculture (where subsistence production is dominant), forestry and fishing.

5. Not only the secondary but also the tertiary sector of the economy is rudimentary, while the infrastructure is skeletal only. This means, to take a single example, that educational opportunities are extremely limited in scale and scope, and the rate of illiteracy is therefore high; probably over half of the population above the age of ten years is illiterate in most of the territories concerned. One result of this situation is that the fortunate few who are able to benefit from such educational facilities as are available (or who can afford education overseas) find themselves in a tremendously powerful position in the social and economic life of the community, while the great illiterate majority have little, if any, real influence in public life and can easily suffer oppression at the hands of the powerful minority.

Differences between the L.D.C.s

So far we have been at some pains to emphasise in what respects the various L.D.C.s resemble one another, but as we shall see later there can be many points of divergence between them. Bauer and Yamey (1965) are concerned to stress the heterogeneity of the underdeveloped world, and they suggest that with regard to the common characteristics which we have already recognised—poverty in personal incomes, poverty in capital and a backward technology—notable differences are discernible on closer examination. Countries with comparatively high growth rates like Taiwan, Thailand and Brazil, for instance, and countries which have already achieved a fair degree of affluence like Venezuela and Singapore fall into quite a different category from territories which remain extremely poor like Mali or Chad, or those which have low, or even negative, rates of growth like Burma, Guyana or Haiti. In material terms the gap between the richer and the poorer L.D.C.s is as great as the gap between the richer L.D.C.s and the developed world: "the richest under-developed country is close to the poorest advanced country." We should, moreover, bear in mind that underdevelopment has by definition nothing to do with such important human attributes as social tradition and culture; India, for example, has a culture and a civilisation which go back for more than 2000 years, though most L.D.C.s in the humid tropics cannot boast of anything comparable to that.

The table below brings out two further points of difference which are of importance to the geographer—differences of area and of population as between selected territories in the humid tropics.

	Area (km^2)	Population	Population density $(per\ km^2)$
Brazil	8,518,477	101,707,000	11·9
Venezuela	912,047	11,000,000	12·1
Haiti	27,749	4,700,000	169
Honduras	111,957	2,800,000	25·0
Ivory Coast	330,223	5,000,000	15·1
Chad	1,261,843	2,755,000	2·18
Zaïre	2,343,940	22,000,000	9·39
Malawi	117,614	4,660,000	39·6
Sri Lanka	65,609	12,748,000	194
Burma	678,031	28,201,000	41·6
Thailand	513,457	34,400,000	67·0
Papua–New Guinea	475,366	2,467,000	5·19

For purposes of comparison we may note that the population densities of the United Kingdom, Belgium and Italy respectively are about 230, 315 and 160 persons per km^2.

The table brings out clearly the two points referred to above. With regard to size there is no comparison between the "pocket" states represented by Haiti on the one hand, and the areal giants such as Brazil and Zaïre on the other, while population densities range from very low in Chad and Papua–New Guinea to very high in Haiti and Sri Lanka. We are clearly dealing with a very wide spectrum of territories in our studies of the humid tropics.

We have so far been considering the general human situation in the L.D.C.s, and we should bear in mind that these cover a very large proportion of the earth's surface. According to Fryer (1958) about 70 per cent of the world's population live in underdeveloped territories, while Shannon (1957) has put the figure at just over 60 per cent. In a later figure Fryer (1965, 13) suggests that 50 per cent of the total world population live in underdeveloped territories and a further 12 per cent in semi-developed economies; only 8 per cent are said to live in highly-developed economies. Some of the L.D.C.s as we have seen lie outside the tropics and we are not directly concerned with them in this book.

One important reason why the inhabitants of the developed countries cannot ignore the problems of the humid tropics is that to an ever-increasing extent they need the resources which many tropical territories have to offer, and the resulting strengthening of world market demand for these products must mean that the peoples of the tropics will not remain as weak a world force as they have been in the past; indeed, there are already clear signs that a marked shift in economic power is already taking place. Ward and Dubos (1972, 288) sum up the situation facing us starkly and pungently, as they ask if there is any kind of sense or justice in the maintenance of a social pyramid with a narrow privileged elite near the apex while at the bottom millions more scratch "a bare living in filthy, workless cities and disintegrating countrysides. If developing peoples were ignorant ... of how the 'other half' lives they might toil on without protest. But the transistor ... and world-wide television have put an end to that kind of ignorance. Can we rationally suppose that they will accept a world 'half slave, half free,' half plunged in consumptive pleasures, half deprived of the bare necessities of life? Can we hope that the protest of the dispossessed will not erupt into local conflict and widening unrest?" This is a question which will become more pressing as time passes, and we neglect it at our peril.

THE PHYSICAL BACKGROUND

It is probably true to say that most people feel instinctively that they know what is meant by the term "the humid tropics." It is no less true to say that close examination of the term quickly demonstrates that we are dealing with a very complex climatic region, as Gourou showed some years ago in his volume *The Tropical World* which was for long the definitive work on this region. Questions of temperature, humidity and rainfall have clearly to be considered in any attempt to arrive at a clear understanding of the actual regions involved. Gourou used a minimum mean monthly temperature of 18° C (64° F) as one of his parameters because he argued that a lower figure could involve the possibility of night frost during the cool season. Even so, slight frost is infrequently experienced in parts of continental West Africa and in North Vietnam, areas which Gourou included within the humid tropics.

Gourou did not place the emphasis upon humidity that some other authorities do as he questioned its direct importance. The point at issue is simply that the presence of humid air at the surface does not of itself guarantee rain; "humid" does not necessarily mean "rainy." This is clearly brought out by Fig. 2, which shows that the humid tropics as broadly defined by Garnier includes such places as Dakar, where the rainy season lasts just over 3 months and where the annual rainfall is only 521 mm (20·5 in.), and—even more surprisingly—Aden, which records a mean annual total of only about 63 mm ($2\frac{1}{2}$ in.) and which has no rainy season worthy of the name. In view of facts like these we can perhaps feel that Gourou is justified in minimising the importance of humidity as a factor in its own right.

Of the importance of rainfall, however, there can be no doubt, but we must bear in mind that it is not only the mean annual total which is important at any given place but also the distribution of rain throughout the year. Fig. 3 will help us on this point. There is clearly a difference in kind, and not merely in degree, between Bombay with its near-2032 mm (80 in.), almost all of which falls in 5 months of the year, and Manaos with 1676 mm (66 in.) and no real dry season at all, though totals in July and August are below 51 mm (2 in.). Even in these months, however, humidity remains high in sharp contrast to the dry months in Bombay. Clearly, vegetational and human responses must differ greatly between these two locations and this fact is generally recognised by assigning them to different climatic and vegetation regions. Different again are stations like Kayes which records a mean annual total of just over 737 mm (29 in.), most of which falls in 5 months of the year, and Freetown with 3500 mm (138 in.), most of which falls in about 8 months, though no month is statistically completely rainless. Each of these stations has a definite dry season, a prolonged and severe one in the case of Kayes.

The mention here of climatic regions reminds us that we have not yet delimited our area of study. We have observed in Fig. 2 Garnier's humid tropics but have seen that the areas included there are not entirely satisfactory for our purpose. Garnier's narrower humid tropics are altogether too restricted while his broader zone in some respects is too

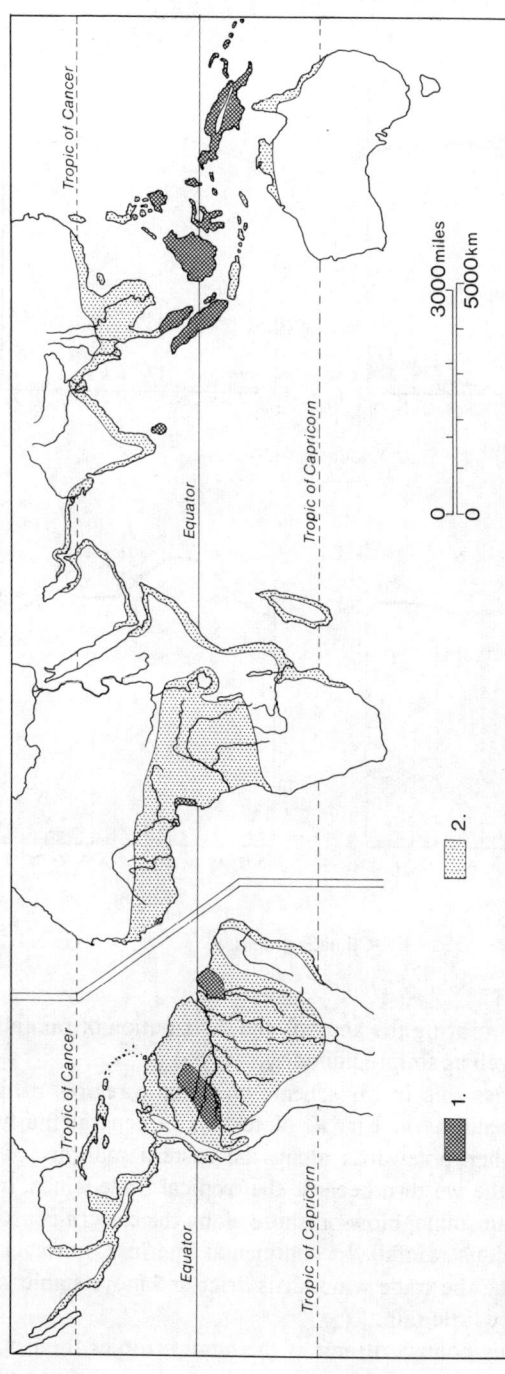

Fig. 2.—The humid tropics.
1. Twelve months with: (a) mean monthly temperatures of 20°C (68°F) or more; (b) a mean annual rainfall of 1016 mm (40 in.) or more; (c) a mean relative humidity of 65% or more; and (d) at least 75 mm (3 in.) of rain in each month.
2. 8–11 months with: (a) mean monthly temperatures of 20°C (68°F) or more; and (b) a mean relative humidity of 65% or more.

[After Garnier]

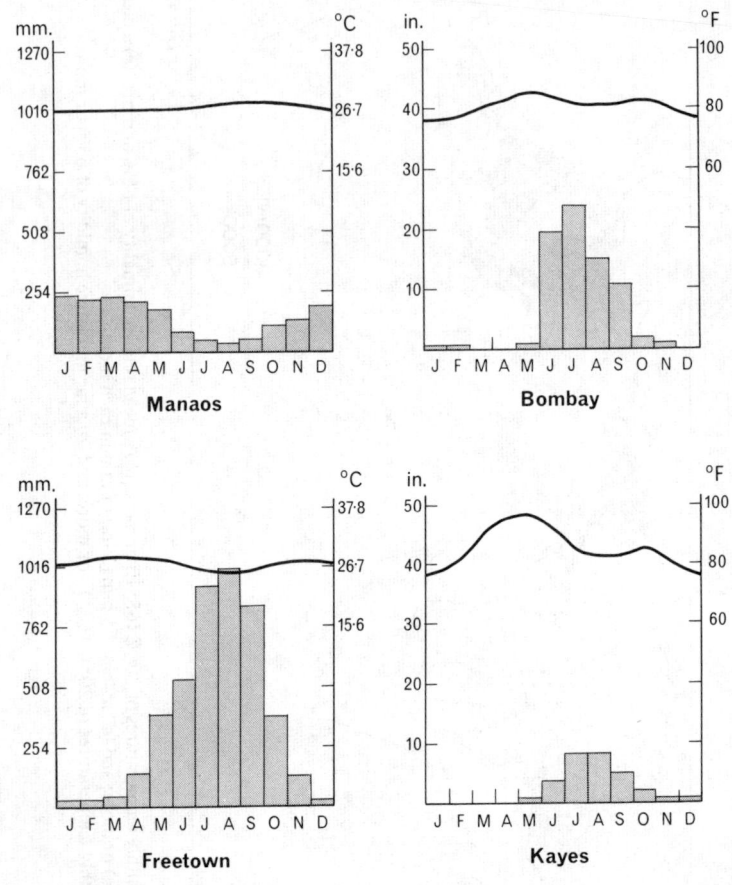

Fig. 3.—Climatic graphs.

extensive. We need to bring the amount and distribution of rainfall into our assessment as well as simply humidity.

Miller (1971) does this in his scheme of climatic regions which is shown, slightly amended, on Fig. 4. Note that in general the humid tropics extend farther polewards along the eastern margins of land masses than along the western because the tropical trade winds, humid and comparatively unstable, blow on-shore along the eastern coasts and so bring fairly copious rainfall. In continental interiors, however, and along western coasts, the trade wind air is drier and more stable, and it therefore brings very little rain.

Miller takes as his poleward limit of the humid tropics the isotherm

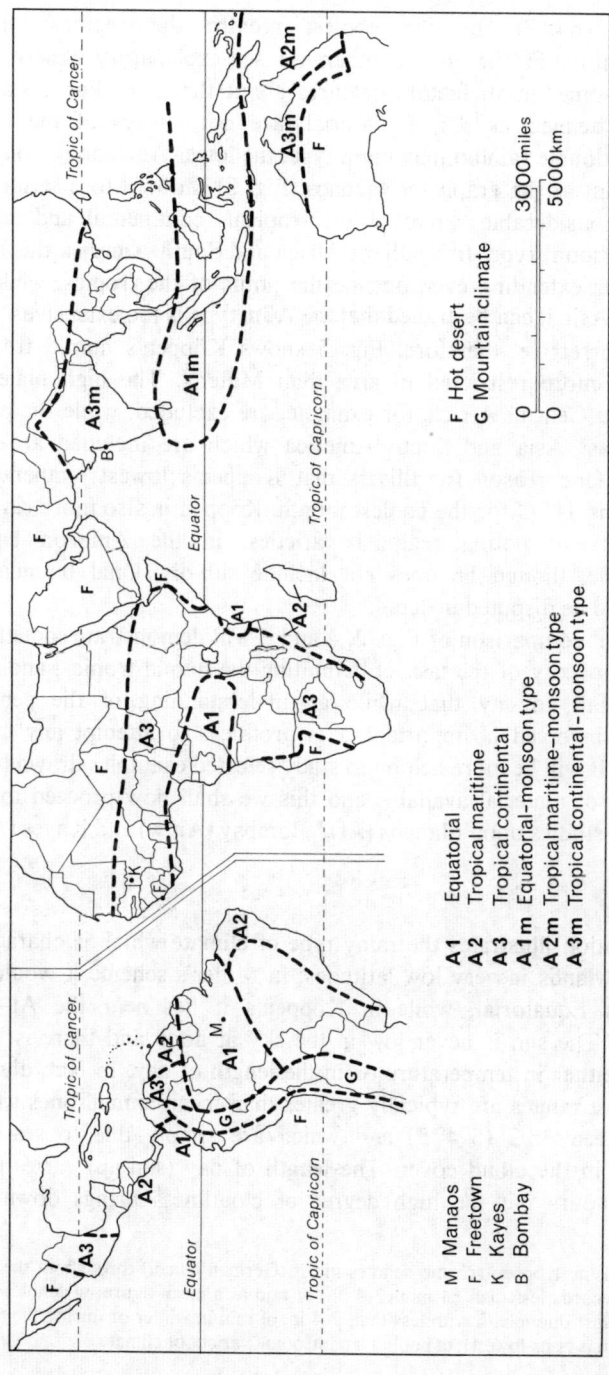

Fig. 4.—The tropical humid climates: 1.

A1 Equatorial
A2 Tropical maritime
A3 Tropical continental
A1m Equatorial-monsoon type
A2m Tropical maritime-monsoon type
A3m Tropical continental-monsoon type

F Hot desert
G Mountain climate

M Manaos
F Freetown
K Kayes
B Bombay

Based partly on Miller

of 18°C (64°F) for the coolest month; the regions which are recognised in Fig. 4 are otherwise self-explanatory. There are, however, some unsatisfactory features about the map. For instance, the areas defined as A1, Equatorial, are not always of the truly equatorial double rainfall maximum type; the lower Amazon region is a case in point as the graph for Manaos (Fig. 3) shows. More serious is the very considerable extent of the tropical (continental and continental monsoon) type. In southern Africa and South America this type is shown as extending even outside the limits of the tropics, while in south-east Asia it can be argued that the A3m type is too extensive.

As a corrective, therefore, Fig. 5 shows Köppen's humid tropics which are more restricted in area than Miller's. The high plateaux of East and Central Africa, for example, are excluded, while the parts of south-east Asia and South America which are included are less extensive. One reason for this is that Köppen's lowest temperature parameter is 18°C for the coolest month. Köppen is also more careful than Miller in noting regional varieties, in the Amazon basin, for example, though he does not include sub-divisional boundaries which could be disputed in detail.

A careful comparison of Figs 2, 4 and 5 will demonstrate something of the complexity of the task of delimiting the humid tropics, and it is probably fair to say that while an understanding of the general principles involved is important it is profitless to attempt any exact definition. It will be more helpful to study selected examples drawn from the wealth of material available and this we shall now proceed to do, basing our studies upon Manaos (Af),* Bombay (Amw) and Kayes (Aw).

Manaos

This station illustrates the rainy type of climate which is characteristic of lowlands in very low latitudes; in Miller's scheme it would be classed as Equatorial, while in Köppen's it lies near the Af–Am boundary. The sun is never low in the sky at noon and there is little variation either in temperature or in the length of day. In fact, diurnal temperature ranges are typically greater than mean annual ones which rarely exceed 3°C (5·4°F) and which are simply due to seasonal variations in the cloud cover. The length of day (sun-up) varies little from 12 hours but the high degree of cloudiness brings down the

---

*The f in the Köppen scheme denotes moist (German *feucht*) throughout the year; no month records less than 61 mm (2·4 in.) of rain as a mean figure: w denotes a dry season (at least one month with less than 2·4 in. of rain in winter or during the period when the sun is at its lowest); m implies a monsoonal variety of climate.

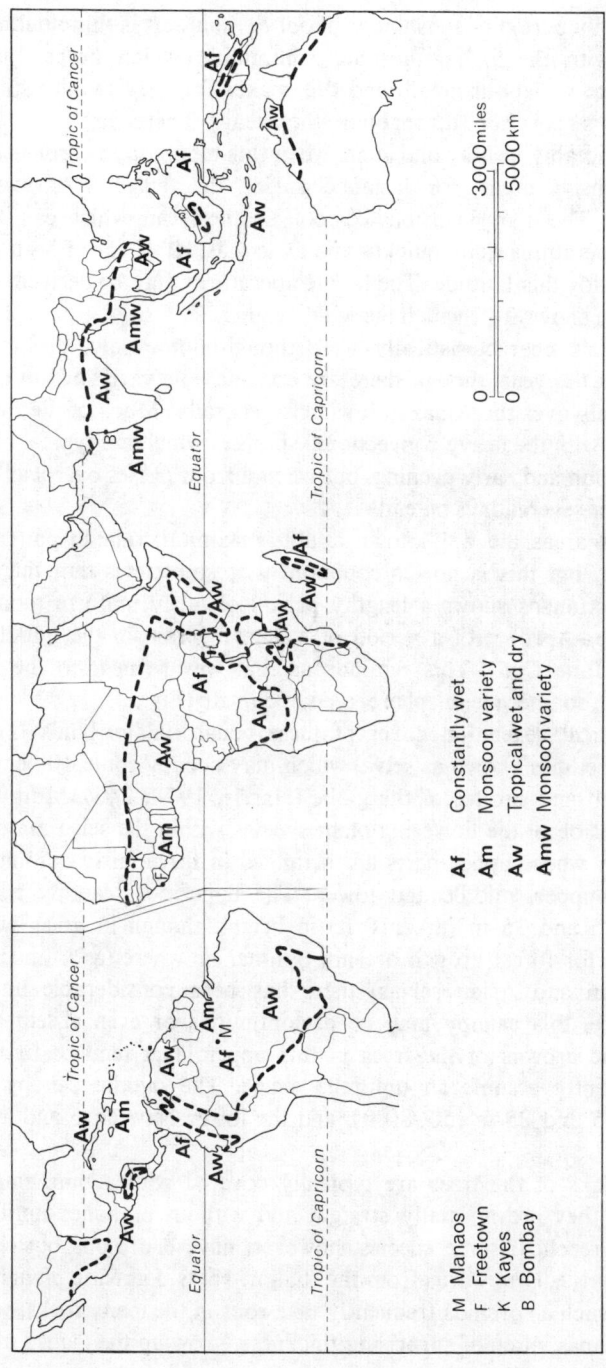

Af Constantly wet
Am Monsoon variety
Aw Tropical wet and dry
Amw Monsoon variety

M Manaos
F Freetown
K Kayes
B Bombay

Fig. 5.—The tropical humid climates: II.

[Based on Köppen]

average daily period of sunshine to about 5½ hours. It is this cloudiness, together with the high atmospheric humidity, which keeps diurnal temperature variations small, and this makes for very humid, stifling, monotonous weather throughout the year. There are occasions, however, notably in May and June, when this monotony is broken over quite extensive areas, for instance during the *friagems* of western Amazonia. These are invasions of cool, southerly air which can bring down temperatures quite quickly and as low as 10°C (50°F)—a very low figure for this latitude. The low temperatures may be derived from the Andean snows over which the wind originates.

Rainfall is characteristically well, though not evenly, distributed throughout the year, though there are considerable variations in mean annual totals over the Amazon lowlands generally. Much of the rain is associated with the heavy convectional showers which are typical of the late afternoon and early evening, but when a front passes overhead rain may last for several days on end.

In some areas the well-known double maximum rainfall pattern is discernible, but this is not as common as is sometimes thought. The graph for Manaos shows a lengthy period of fairly uniform monthly totals (Dec.–April) with a period of slackening rain in the middle of the year (June–Oct.). This reminds us that we are near to the Am, monsoonal, southern hemisphere, type of climate (Fig. 5).

The typical vegetation cover of these equatorial lowlands is rain forest, sometimes known as selva, which may occupy some 10 per cent of the total land surface of the globe (Harris, 1974, 243). More than 90 per cent of all the flowering plant species within the selva are evergreen trees whose upper parts are arranged in three fairly continuous canopies—upper, middle and lower. The uppermost canopy ranges between 25 and 35 m (80–110 ft) in height, though in areas where conditions for forest growth are not optimal or where (as in much of the African and Asian selvas) there has been considerable human interference, this canopy may be discontinuous or even absent altogether. The crowns of the trees in this upper layer tend to be wide and frequently assume an umbrella shape. The middle canopy lies between 15 and 25 m (50–80 ft), and the lower between 5 and 15 m (15–50 ft).

The trunks of the trees are typically covered with a thin, smooth bark, and they are normally straight and without branches until the canopy is reached; some species, however, have enormous buttresses growing vertically upwards from the shallow roots. Parasitic plants and epiphytes such as orchids frequently take root in the many forks, while vine-like lianas, often of surprising thickness, grow up the trunks of the main trees or hang from the branches, frequently forming festoons between neighbouring trees. The interior of the forest is gloomy and oppressive for the air is still and humidity high. Even at midday the high sun cannot penetrate through the thick canopies above and a kind of unnatural twilight reigns. Vegetation on the ground is sparse and is largely limited to various forms of fungi and ferns.

It is often argued that tropical rain forest provides a very real barrier to human penetration and settlement but this is not normally the case as undergrowth, as we have noted, is generally lacking, though it is true that the presence of slippery clay soils and fallen branches and trees can greatly impede progress. On the whole, however, it is lack of incentive which has thrown up the barrier to progress in the selva regions, not the rain forest itself. Where the forest has been cleared, however, a very different situation rapidly develops as the sunlight can then penetrate to ground level; rapid vegetational growth and regeneration takes place under these circumstances, and dense, often impenetrable, undergrowth develops as it does along river banks where the sunlight also reaches the ground. It is estimated that after the clearance the development back to the true primary forest takes about 250 years if no further interference takes place—as in fact it frequently does.

The tropical rain forests are remarkable for the large numbers of plant species which they include. It is said, for instance, that in Brazil alone there are about 40,000 different species of flowering plants, and in the state of Pará alone 2500 different species of trees have been counted. It is quite usual for as many as 40 different tree species to establish themselves in a single hectare, and sometimes the figure may rise to 100. This contrasts markedly with the corresponding figure of fewer than 10 species per hectare in many parts of the temperate deciduous forests. The same point is emphasised by Gourou (1968) who states that, for instance, among 100 trees growing next to each other in Madagascar there will typically be up to 40 different species. In the Ivory Coast there are at least 500 species of trees while on Mount Makiling, a small volcano on the island of Luzon, there are more species of woody plants than in the entire U.S.A. It is unusual to find single species stands in the rain forests except in individual cases such as the teak forests of Burma or the *Mora* forest of Trinidad.

It is important to remember that quite considerable areas within the rain forest region do not, in fact, support selva. In some places, for instance, very extensive swamps inhibit rain forest growth, while water surfaces can cover broad areas. Low-lying coastal areas frequently support mangrove swamps, while sandbars and mudflats and other areas

of newly exposed rock may be occupied by seral plant communities.*
In some areas unfavourable conditions of relief or soil may inhibit the
development of rain forest and give rise to forest of a degraded type,
while forest areas which have been repeatedly cleared and burnt may
become areas of grassland, as has happened over much of south-eastern
Asia; almost one-fifth of the Philippines is now grassland for this reason.

The most extensive area of rain forest is, of course, that occupying
the Amazon lowlands in South America, and this extends in an
unbroken cover from north to south between the Guiana and Brazilian
Plateaux, and from west to east between the Andes and the Atlantic
coast. Narrow extensions occur along the Atlantic coast southwards
almost as far as 30 degrees south, while along the Pacific coast it
extends northwards from the Gulf of Guayaquil to Panama where it
covers the whole isthmus before extending farther northwards into
southern Mexico. Most of the Caribbean islands are also forested,
though not entirely with true selva.

In south-eastern Asia rain forest covers most of the East Indian
islands. A narrow strip occurs in Australia along the coasts of northern
Queensland, Northern Territory and a small part of Western Australia,
while to the north it covers the Malayan Peninsula and extends as far
as Burma, southern Bangladesh, and even into south-east China though
it merges into monsoon forest along its northern boundaries. The
African rain forest area is the smallest of the major groups; it covers the
northern parts of the Zaïre basin and extends westwards into southern
Nigeria. A detached section extends from south-western Ghana as far
as the Republic of Guinea.

* If bare rock is freshly exposed as a result, for example, of volcanic action, mass movement or deposition, the first plants to appear on the new surface will be lichens and algae, followed by mosses, various kinds of grasses, shrubs and finally trees—small ones at first and then of increasing size. The vegetation becomes more complex as earlier and simpler forms of plant life create conditions suitable for more highly-developed species. As time goes on more plants colonise any given area and the early open conditions are replaced by a closed vegetation cover. A "primary succession" such as has been outlined needs a very long period of time before it reaches its final stages, largely because of the slowness of soil formation.

The succession of plant communities which colonises any given site from the earliest to the final stage is known as a *sere*; the intermediate vegetational groups are known as *seral communities*, and the final community which continues indefinitely is the *climax*. A sere is quite different from the secondary growth which normally moves in to re-colonise land cleared of its former vegetation but which is likely to possess a soil covering in which seeds from the earlier phase are likely to lie embedded. Grasses and shrubs, and even small trees can establish themselves comparatively rapidly under these conditions, and the earliest stages of seral communities will not appear at all. Under humid tropical conditions an abandoned clearing is likely to be thickly vegetated by shrubs and small trees within 4 or 5 years, though the development of the true climax vegetation again will take very much longer.

Bombay

This station is characteristic of Miller's tropical monsoon type of climate and Köppen's Amw, and it shows marked differences from the previous type. Temperatures, for instance, show greater variations, the mean annual range being greater; thus the figure for Bombay itself is 4·5° C (10° F), while in some interior areas it is as much as 8° C (15° F). Rainfall is seasonal and, moreover, the rainy season begins with a rush; it is not without reason that the Indian monsoon is said to "burst." This feature is suggested by Fig. 3, for while May records a mean figure of just under 20 mm (less than 1 in.) that for June is about 510 mm (about 20 in.).

It is sometimes suggested that four seasons can be recognised in the main (south-east Asian) monsoonal region.

    1. December–March. This is the cool season when the sun is comparatively low at noon. Winds are predominantly north-westerly (the winter monsoon) and are therefore off-shore—with a few exceptions due to coastal direction. Rainfall is very low except in some areas of on-shore winds.

    2. April–May. This is the hot season when the sun is high and temperatures soar. Relative humidities also increase though there is little cloud, and conditions therefore become extremely trying. Storms occur in some areas but they bring little relief from the stifling heat.

    3. June–September. The south-west monsoon breaks through in June and rainfall immediately is heavy; there is no gradual increase as there normally is in regions of tropical (Aw) climate. Relative humidities remain high but temperatures fall owing to the heavy cloud and rain, and living conditions as a result become much more bearable. The monsoon air stream does not move as a homogeneous mass; surges occur at intervals and these result in local fronts and convergences. Along these frontal zones occur strong winds, thick cumulus cloud and very heavy rain. It has been said that it can be dangerous to be out of doors during periods of the heaviest rain as there is so much falling water in the atmosphere that it can be difficult to breathe!

    4. October–November. This is sometimes known as the season of the retreating monsoon, and the south-west monsoon is much lighter during this season as its northern limit gradually recedes southwards. Rainfall is much diminished but relative humidities remain high and temperatures increase. Conditions are therefore

trying at first but they gradually improve as the cool season approaches.

The typical vegetation of the tropical monsoon lands is forest, but of a different type from the selvas. It is sometimes known as semi-deciduous forest but its origin is open to some doubt because in parts of India and Burma where population is comparatively low and human interference therefore minimal the forests frequently assume the appearance of true rain forests. It may therefore be inferred that the semi-deciduous forest, characteristic though it is of extensive areas, is to some degree a response to human as well as to climatic factors.

The morphology of the semi-deciduous forest is easily described. An upper canopy at a height of between 18 and 24 m (60 and 70 ft)—roughly the height of the middle canopy of the rain forest—lies over a lower canopy at about 12 m (40 ft). The higher trees are deciduous and in drier areas they become fairly widely spaced, while the lower ones are evergreen. The trees are normally more widely spaced than in the true rain forest while they have fewer buttresses but more lower branches; they also have thicker trunks and bark which retain moisture during the dry season. It is important to notice that many parts of the tropical monsoon lands, notably in Asia, include areas which have been settled for a very long time indeed and which are today very densely populated, and as we have already noted an inevitable result of this is that the forests in these areas have been greatly modified by human agency.

The semi-deciduous forests of the world are most widespread in south-eastern Asia where they are customarily known as monsoon forests. They occur particularly in India, Bangladesh, Burma and the northern parts of Thailand, Laos and Vietnam. In the western hemisphere they form a discontinuous fringe around the rain forest and they cover extensive parts of the West Indies. They are rare in Africa.

Kayes and Freetown

Kayes is representative of the tropical (wet-and-dry) type of climate, Köppen's Aw and Miller's tropical continental types, which occurs between the equatorial regions on the one hand and the arid zones of the tropics and sub-tropics (the hot deserts) on the other. *Freetown* represents a wet variety of this climatic type which Köppen designates Am. A dry season coincides with the period when the sun is near its lowest noon elevation and while it is beginning to strengthen, while a rainy season occurs when the sun is near its highest point and when it is beginning to retreat. This seasonal change is closely linked with latitudinal movements of the rain-inducing inter-tropical convergence zone which follows at some distance in space and time the northward and southward movements of the sun. The dry season therefore lengthens with increasing distance from the equator, varying from about 2 months along the equatorial margins to 5 or 6 months near the drier boundaries. Freetown and Kayes respectively (Fig. 3) illustrate this.

There is a noticeable similarity between this climatic type and the tropical monsoon, though as we have seen there are also differences. Temperatures on the whole follow a comparable pattern though, particularly in continental interiors, the annual range of temperature is considerably greater and may exceed 11°C (20°F). In maritime areas it is much less, sometimes no more than 3°C (5°F). Rainfall in general is less than in equatorial or monsoon regions though considerable variations occur. As a general rule amounts diminish with increasing distance from the equator though aspect and relief frequently modify this overall pattern. A mean annual total of 610 mm (24 in.) is sometimes accepted as the minimum for the humid tropics though this point is debatable.

The march of the seasons throughout the year is closely parallel to that experienced in the tropical monsoon regions and this need not be repeated. Notice, however, that the rainy season does not open as suddenly as it does under truly monsoonal conditions. The rains begin more gradually with storms of the line squall type, and these become more intensive and more frequent until the period of prolonged rain sets in. Similar storms mark the close of the rains.

The natural vegetation of the wet-and-dry tropics has traditionally been recognised in text-books as being of the type known as savanna, a word which comes from an Amerindian word used in parts of the Caribbean to denote a treeless grassy plain. The use of the word has now been broadened to include other areas of tropical grassland which may comprise various combinations of grass, shrubs and trees. Local variations are very considerable and this fact gives rise to serious doubts regarding the exact status of the savanna type of vegetation. Schimper (1903) believed that savanna grasslands form a true vegetation climax of the wet-and-dry tropics, but since his work was published it has become increasingly clear that most savannas (with the possible exception of some in Australia) show the effects of relief, drainage, soil, fire, and frequently of human interference, for it now seems likely that they occur in regions which would naturally develop towards a woodland vegetational climax.

Figure 6 will help us to understand this. It so happens that the greater part of the African and Brazilian savannas lie on ancient shields which have been uplifted and which show a surface relief of upland

```
       Sw      D  Sg  D      Sw      D  Sg  D      Sw           E
```

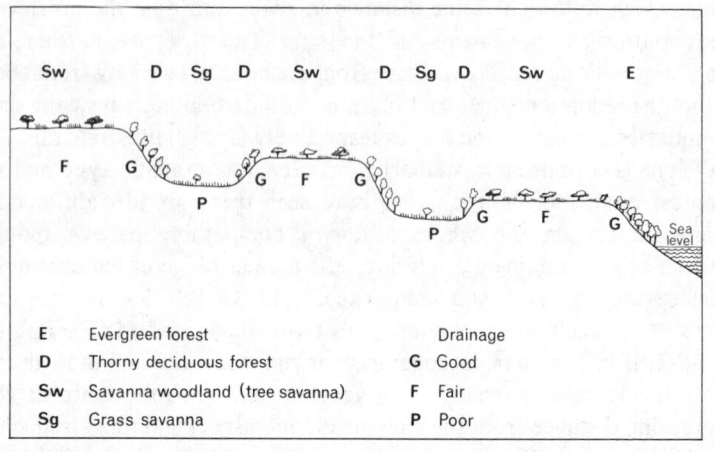

E	Evergreen forest		Drainage
D	Thorny deciduous forest	**G**	Good
Sw	Savanna woodland (tree savanna)	**F**	Fair
Sg	Grass savanna	**P**	Poor

Fig. 6.—The relation between plant communities and relief in the wet-and-dry tropics: I.

erosion surfaces (the pediplanes of many writers). On the lower surfaces such as those shown in Fig. 6 the soils frequently consist of impermeable clays or sandy residues overlying mottled clay or ironpan. Drainage will therefore be very poor because water cannot easily soak downwards neither can it freely move laterally over the generally flat surface. While the soils may become parched and hard during the dry season, during the rains water occupies every depression, and after heavy storms may remain standing over much of the whole surface area. True forest growth shuns such a location which normally supports grass savanna, while savanna woodlands or tree savanna* are more usual on the higher, slightly better-drained pediplanes. Forests occur on sloping but well-watered areas and are either deciduous or evergreen according to the amounts of water received, but an interesting contrast to the rain forest is that comparatively few species of trees are found; indeed, the dominance of a single species is by no means unknown as in the acacia woodlands of East Africa. The main points at issue are shown schematically on Fig. 6, while Fig. 7 shows another possible situation. Evergreen forest occupies well-watered but well-drained slopes, scrub or grass savanna occurs on the areas of poorest drainage, and scrub or savanna woodland where drainage improves.

It is an important point that the grasses which form such an important feature of savanna landscapes are quite different in nature from those of temperate latitudes. They grow from swollen roots such as rhizomes, bulbs or tubers, and they have a tussocky form; these are

* The trees stand well apart in the tree savanna but not in the savanna woodland.

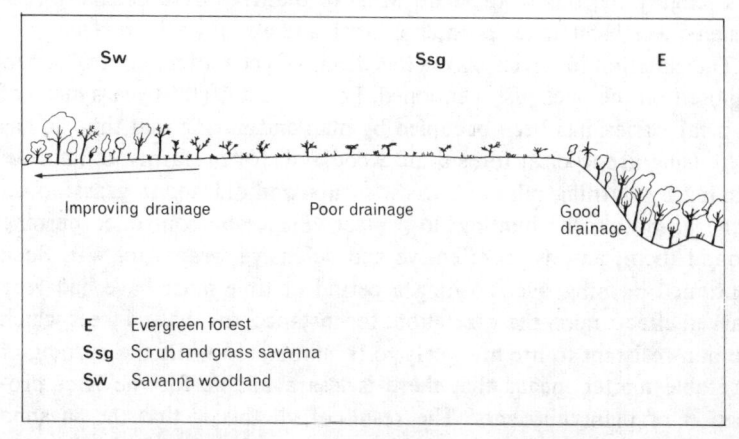

FIG. 7.—The relation between plant communities and relief in the wet-and-dry tropics: II.

Diagram labels: Sw (Improving drainage), Ssg (Poor drainage), E (Good drainage).

E   Evergreen forest
Ssg Scrub and grass savanna
Sw Savanna woodland

features which help to guard against destruction by fire. The swollen roots are not as susceptible to fire as "normal" grass roots, while the tussocks protect the inner core of the plant and thus allow regeneration after all but the fiercest fires. The grasses are usually coarse and rough except in their earliest stages of growth, and they have broadish, flat leaves. They grow to heights of at least 75 cm even in the drier areas, while the well-known elephant grass can form an exceedingly dense thicket up to 5 m in height.

Enough has probably been said to show that relief, soil and drainage all act as powerful controls in the development of savanna landscapes, while another powerful control is fire which frequently breaks out at the beginning of the rainy season when the vegetation is still very dry. It is started by the lightning which accompanies the storms which form a prelude to the true rainy season and they probably play a substantial part in modifying the vegetation cover. These fires, of course, are of natural occurrence, unlike those begun by farmers towards the end of the dry season.

Savanna trees are mainly deciduous and fire-tolerant, and are rarely more than 12 m (36 ft) high except where they are associated with relict patches of true forest. Many of the trees and shrubs are thorny, with fire-resisting bark and small leaves which minimise the rate of transpiration, while many are capable of regeneration from dormant buds and suckers after a fire. Some, notably the baobab, have very thick bark and a spongy type of wood which can hold large quantities of water, and these features all increase their ability to withstand both fire and lengthy periods of drought. Most of the trees have extensive root systems which can take up water at depth and also near the surface.

The situation in Africa shows the effects of yet another control superimposed on the ones just mentioned. For at least 10,000 years much of tropical Africa has been occupied by man and during that time he has been using fire to clear forests and woodlands for his farms; to improve grazing by burning off small trees, shrubs and old, tough grass; to set game in motion for hunting; to protect villages by controlled burning around them; and as an offensive and defensive weapon of war. Such continued burning over so long a period of time must have had very marked effects upon the vegetation; for instance, shrubs and trees which are non-resistant to fire are likely to be eliminated while the burning of vegetable matter means that there is less available for the later production of plant nutrients. The result of all this is that the savanna slowly suffers a steady degradation. Forests of various types become savanna woodland or even grass savanna, while savanna woodland frequently becomes tree savanna. Relics of the former richer vegetation often remain along rivers and watercourses and sometimes as scattered clumps.

Savanna lands are very extensive, and they may be increasing in area. In fact, they may account for about 13 per cent of the total land surface of the globe. There are only small patches of savanna in Central America but extensive areas occur in South America in the *llanos* of the Orinoco basin and the *campos* of central and south-western Brazil. In Africa they extend right across the continent south of the Sahara Desert and they cover large parts of the plateaux of East and Central Africa. In Australia they occur to the south of the tropical rain forests in the north of the continent.

In conclusion, it might be useful to note that savannas do not normally directly border rain forest. Where the average length of the dry season exceeds $2\frac{1}{2}$ months the rain forest begins to change to seasonal forest and woodland, while the woodland in its turn gives place to savanna. There is a progressive shift as the dry season lengthens from evergreen to deciduous forest and a corresponding reduction in the amount of tree cover. Beard (1955) recognises three transitional types between the two extremes; semi-evergreen seasonal forest, deciduous seasonal forest, and thorn woodland. The first type is widespread in the American tropics particularly in the *cerradão* of Brazil, while it also occurs in West Africa (the dry evergreen forest) and in south-east Asia (the wetter monsoon forest). Trees tend to be rather less tall than in the rain forest, and buttress growth is uncommon, while they have thicker bark. Epiphytes and lianas are common but they include species which need less moisture than those of the rain forest. Surface vegetation (shrubs and small trees) is better developed than in the selva as the more open canopies allow more sunlight to reach the ground.

Deciduous trees are naturally dominant in the deciduous seasonal forests and they tend to have umbrella-shaped or plate-like crowns. They, too, typically have thick bark, small leaves and extensive root systems, while lianas and epiphytes are few. In Africa this forest type is often known as savanna woodland or *miombo*, while in south-east Asia it corresponds to the drier monsoon forest. Thorn woodland is most widespread in the American tropics, particularly in the *caatinga* of the dry zone of north-eastern Brazil.

FURTHER CLIMATIC CONSIDERATIONS

We have gone into some detail with regard to climate and vegetation in the humid tropics in an attempt to show that these regions are by no means as uniform throughout as is sometimes believed; very considerable variations exist, and these are at least as pronounced as the corresponding variations which occur in temperate lands. It is grossly inaccurate to imagine that the humid tropics forms one vast fairly homogeneous region characterised by great heat, humidity and rainfall. Our studies of the vegetation alone show that many factors must be taken into account in order to explain the variations which occur, and this is a very important point, for such factors will also play leading parts in determining the success or failure of cultivated plants.

We have, however, far from exhausted the list of climatic variations which must be considered if we wish seriously to consider the value of the humid tropics for human occupance and use. We might at this point examine one or two features which have hitherto been omitted.

1. *Local rainfall anomalies.* Rainfall anomalies are of frequent occurrence throughout the humid tropics. Examples include the well-known dry areas of northern Colombia, northern Venezuela and the north-eastern shoulder of Brazil in South America; the coastal zone between south-eastern Ghana and south-western Nigeria in West Africa; and central Burma in Asia. These anomalies are usually explicable in terms of relief, ocean currents or the vertical structure of the atmosphere, but they present special problems from the point of view of land use.

2. *Annual rainfall variations.* It is a well-marked feature of tropical climates that at any given place rainfall frequently varies

considerably from year to year. Consider the following figures for Banjul, Gambia (figures in mm and in.).

1901	1151	(45·31)	1910	1118	(44·00)
1902	747	(29·42)	1911	715	(28·14)
1903	1450	(57·13)	1912	863	(33·99)
1904	966	(38·02)	1913	601	(23·68)
1905	1678	(66·07)	1914	1242	(48·91)
1906	1635	(64·36)	1915	1210	(47·64)
1907	864	(34·00)	1916	966	(38·02)
1908	1106	(43·54)	1917	957	(37·68)
1909	1437	(56·59)	1918	1373	(54·03)

Variations of this magnitude are of tremendous importance for farmers, and indeed for all the inhabitants because the success or failure of food crops largely depend upon them, but unfortunately they are entirely random and cannot be predicted in advance. The average of the above set of figures works out at a mean annual total of about 1092 mm (43 in.) but this figure is of much less significance than actual totals which vary between 601 mm (23·68 in.) and 1678 mm (66·07 in.). There are indeed sequences when rainfall can be below average for 3 years running (1911–1913) and this can be extremely serious for the inhabitants.

Closely associated with the above is the way in which similar variations occur from year to year on a monthly basis. This is shown by the following figures, also for Banjul, which show recorded maxima and minima totals in the "rice months." Figures are again in mm and in.

	Maximum		Minimum	
May	48	(1·9)	0	(0·0)
June	313	(12·32)	56	(2·24)
July	354	(13·93)	130	(5·10)
August	497	(19·56)	173	(6·79)
September	332	(13·08)	145	(5·78)
October	230	(9·08)	6	(0·24)
November	0·5	(0·2)	0	(0·0)

The above figures, of course, do not refer to any single year.

Difficulties experienced by cultivators are highlighted by these figures as it is often very difficult for farmers to know just when to sow and when to reap. In some years, for instance, June has plenty of rain to nurture the germinating and newly-growing crop, but in other years it has not. In some years it would be perfectly safe to begin harvesting in October as the rains have virtually finished—but not in other years. These are very practical points which are of far more concern than mean or average rainfall figures.

3. Another important characteristic of tropical rainfall is its *intensity*, though the highest intensities are only normal over short periods of time, probably for less than 1 hour. Even so, this is a very real factor to be reckoned with. Dakar, for instance, has recorded 89 mm ($3\frac{1}{2}$ in.) of rain in a single storm which lasted for 72 minutes, while Freetown has experienced 152 mm (6 in.) in an hour and 305 mm (12 in.) in 1 day. In Malaya, rainfall intensity often reaches 150 mm (6 in.) in an hour while a study of storms in Java showed that 22 per cent of the total annual rainfall originated in cloudbursts, which were defined as storms with an intensity of 1 mm per minute for at least 5 minutes.

Rainfall on this scale is not just inconvenient; it is highly dangerous. A very large proportion of a season's rainfall can be concentrated in heavy storms and more water can therefore be lost in run-off than is conserved by seepage, and it does not, under these circumstances, become available for agriculture. In addition the violent run-off can produce serious soil erosion, while surges of flood water can wreak very considerable havoc among crops standing in their path.

4. *Temperatures* in the humid tropics are less extreme that is sometimes believed, though as we have seen earlier the range increases in continental interiors. Even so, maximum temperatures throughout the humid tropics are significantly lower than those in the dry tropics or sub-tropics where the greatest extremes occur. In general, however, temperatures in the humid tropics are at all times high enough to permit plant growth; it is rainfall which is the regulator.

5. The *consistently high relative humidities* which are so typical of the humid tropics can produce very trying living conditions and considerably reduce the output of human energy. Indirectly, as they encourage the growth of maleficent organisms, they encourage disease and so endanger health unless strict precautions are taken. It is worth noting that during the rainy season relative humidities frequently approach or attain 100 per cent in the early morning, though they fall somewhat later in the day as temperatures rise.

6. Although in general *winds* are light throughout the humid tropics (this, oddly enough, is contrary to popular belief though it has long been recognised that the doldrums form an area of calm near the equator), strong winds do occur in most areas from time to time, and these can be extremely violent and destructive. It is for the most part the peripheries of the region which suffer most from destructive storms, examples including the West Indies, Central America, Madagascar and the nearby coasts of East Africa, the

Philippines, and the coasts of south-east Asia. The *sumatras* of southern Malaysia and Singapore and the line squalls of West Africa can also be very destructive.

On the other hand, some winds are beneficial in some measure at least to the areas over which they blow. A wind of this type is the *harmattan*, the air stream which affects most of West Africa during the dry season and which is the normal trade wind strengthened by the North African cool season zone of high pressure. It advances southwards as the rains retreat during the last quarter of the year and retreats northwards the following year as the rains advance. At its greatest extent it can even reach the Guinea Coast but for very short periods only. It is, of course, a very dry wind, usually cool though sometimes hot. The strongly desiccating effect which characterises it exercises a strong influence upon crop distribution; tree crops, for instance, like coffee or cocoa cannot withstand it for anything but very short periods. Over the northern parts of West Africa it brings very trying conditions because of its extreme dryness, but farther south it is greatly valued in terms of human comfort as it brings welcome relief from the general high humidities—hence its affectionate name "the doctor." Its extreme dryness is considerably ameliorated in the southern Guinea lands, but in the north it causes great discomfort as the skin dries out and frequently becomes chapped, while cracks appear and produce painful sores in sensitive areas around the nose and lips.

The most notable contribution which the *harmattan* makes to West Africa, however, lies in the enormous masses of fine sand and dust which it sweeps southwards from the Sahara and deposits over almost the whole region, though, of course, in diminishing amounts southwards. It is now generally agreed that these deposits make a most valuable contribution to the fertility of many West African soils and agricultural production would be much diminished without them.

Chapter II

Natural Resources of the Humid Tropics

THE ROLE OF THE ENVIRONMENT

IT is by no means an easy task to assess accurately the role of natural resources in economic development. This may appear surprising since it is clear enough that such resources are indispensable because without them there could be no development at all—a fact which geographers have been quick to recognise. Many writers, however, have evoked criticism by their tendency to discuss at length the various "physical" (or "geographical"; significantly the two terms have frequently been viewed as virtually synonymous) factors involved and to ignore the "human" (including economic) side of the equation altogether. This line of approach was characteristic, for example, of Ellen Semple (1911) who went so far as to argue that if another world could be constructed with exactly the same physical basis and therefore the same natural resources as our present one, the course of human history on the new planet would parallel exactly that of the old.

While few, if any, geographers would go as far as this today there is little doubt that this preoccupation with "geographical factors" has greatly lessened the impact made by geographers in public realms of policy and planning. There is an instinctive feeling that such an approach is wrong, and many thinkers, perhaps recalling simplistic attempts made by school geography teachers to explain various complex human phenomena in terms of their physical environment, reject it. It is just possible that in the lower forms of secondary schools there is something to be said for this deterministic approach for the simpler physical factors are comprehended fairly easily while human aspects of a case are frequently more difficult to understand, and it may be that at this level teachers are correct not to confuse the issue in the minds of their pupils, but there is nothing whatever to be said in its favour at higher levels. Unfortunately, it is still the case that in some examinations and in some text-books geography students are urged to account for spatial phenomena in terms of supposedly relevant geographical factors. Three comments may perhaps be made at this stage.

1. The term "geographical factor" is a very unsatisfactory one. The present writer (Jarrett, 1974b, 19) has referred to this, pointing out that the meaning of the term is usually narrowed down to include only factors which in themselves are of a geographical nature, and this in practice normally means relief, geology and structure, soils, climate and mineral deposits. Such a basis for study is clearly far too circumscribed, and we should also be concerned with factors which, whatever their intrinsic nature, have *geographical repercussions*— a much broader area of enquiry.

2. An environment (even a physical environment) is not a static feature. It varies from time to time in accordance with variations in other factors. We may take an obvious example from Britain where natural resources such as coal, natural gas and mineral oil have remained in existence beneath the land surface and the continental shelf ever since the present crustal configuration of the region came into being, but it is only comparatively recently that these minerals have become part of our *human* environment. Even as late as 1660 the estimated coal production in England was barely $2\frac{1}{4}$ million tonnes *per annum*—about 400 kg per head of the total population—and it was not until coal became more widely used in industrial processes in the eighteenth century that the large-scale development of the coalfields began. And we have had to wait until our own day to see even the beginnings of the development of the natural gas and mineral oil which lie beneath our continental shelf. The mere existence of these resources is one thing; it is an entirely different matter to arrive at such a stage of development that we can extract them and make use of them so that we can say that they form part of our environment.

3. The state of the world market is continually changing, a point connected with 2 above. Not only must the technological means essential to production exist, but the strength of the market must be such that production becomes worth while. In the early 1960s, to take one example, there were sufficient nickel supplies to meet all world demands, but between 1963 and 1966 demand rose steeply at a rate of over 20 per cent *per annum*. By the end of 1966, therefore, supplies were quite insufficient, and a critical shortage had developed—ironically enough, a shortage of a metal which is the fifth most common element in the earth's crust! This situation came about because of the sharply increasing use of alloys in which nickel plays a leading role. Additional reserves of nickel present in commercial quantities have been discovered since that date, however, and territories in the humid tropics which are benefiting from this expansion of the market include Guatemala and New Caledonia.

THE PRESENT SITUATION

Old habits and modes of thought die hard, however, and as Wagner (1960, 5) has observed the purging of "environmentalist" patterns of thinking from academic geography has been a very lengthy and at times painful process. We may doubt, in fact, whether even now the purging is complete though it is certainly well advanced. Hodder (1973, 8) draws our attention to the fact that many authorities—so far has the pendulum now swung in some quarters—would go so far as to disregard entirely the role of natural and physical resources in economic development. They would argue that economic growth is dependent upon capital—"the engine of growth"—and that natural factors may be disregarded.

Take, for instance, the natural resource we call "land." Certain of the classical economists, notably Ricardo and Malthus, agreed that land was fixed in amount and they believed that this fact alone would severely limit economic growth and population increase. But these early ideas were apparently disposed of by subsequent experience as new lands for development were discovered in Australia, South Africa, North America and elsewhere, and the flood of cheap food which the opening up of these lands released seemed to dispel the Malthusian gloom. Trade increased by leaps and bounds and the natural resource limitations of the industrialised world were made good by the interchange of commodities between different countries. It was under the influence of this apparent economy of plenty that the feeling came about that the role of natural resources could be ignored because the natural development of trade would serve to carry essential commodities to the regions where capital was available to make use of them.

It is significant that today this "economy of plenty" philosophy is under fire from all quarters as it becomes progressively clearer that it flies in the face of contemporary experience. We are re-learning the fact that even global resources are not limitless though the lesson is neither easy to learn nor greatly to our liking. It may be, therefore, that we shall in the future hear more about the role of natural resources than in the recent past, and in fact signs are not wanting that this is already happening. It does indeed seem to be a matter of common sense that all available resources existing within a region should be considered as (potentially or actually) making their individual contributions towards development. It may be convenient to distinguish between natural and human resources for purposes of analysis but in fact they both form part of the total environmental symbiosis and they should be viewed in this light.

Commentators are not lacking to warn us of shoals and reefs ahead. Among them is Lord Ashby, Master of Clare College, Cambridge and Chairman of the High Commission on Environmental Pollution between 1970 and 1973, who (*The Times*, 19th December 1974) has commented upon the increasing shortages now being felt of many resources, and has shown how producing nations have gained from this. The Arabs have quadrupled the price of oil; Morocco in 1973 tripled the price of phosphates; and Jamaica has acted in a similar manner with respect to bauxite. The costs of American and Canadian wheat have risen relatively as much as the cost of oil. These matters are of the greatest importance to all nations, and they press home the point that the days of the economy of plenty are over.

RESOURCES AND THE ENVIRONMENT

It is an undisputed fact that the distribution of natural resources over the earth's surface is very uneven—a matter of great importance to geographers and to planners. It is therefore by no means irrelevant that considerable importance is paid in this book to the study of resources, and two points may usefully be made at this stage.

1. There are different "levels" at which the study of natural resources has relevance and importance, the two extremes being the global and the local levels. The global strategist attempts to relate the world distribution of particular resources to the areas of demand, and he is interested in studying the world-wide developments of trade which accompany the attempts made to match global supply and demand. This line of study must involve consideration of the effects of inter-territorial movements of capital and international relationships, for these matters intimately affect trading patterns, but interesting and important as it is when it gets away from the arid "commercial geography" of former years it does not lie at the centre of our present sphere of interest.

It is rather at the local level that our present study is for the most part pitched. The development of natural resources takes place within the confines of individual territories and it is with the developing territories of the humid tropics that we are primarily concerned. We need therefore to take note not only of the varying distribution of available resources, but also of the contributions which these resources can make to the individual territories concerned. It is, for instance, a relevant point that in the case of small territories (and some of the territories of the humid tropics are very small) comparatively limited resources are able to make a substantial contribution to their economic and social progress though at the global level the resources concerned may be of limited impact. The bauxite reserves of Jamaica and the Republic of Guinea, for example, come into this category, as do the manganese of Gabon and the mineral oil of Brunei, while in the agricultural sector a comparable situation exists with regard to the production of groundnuts in the Gambia and in Senegal. On the other hand, large territories like Brazil and Zaïre need large-scale resources markedly to affect their rates of development.

2. We must beware of falling into the old deterministic trap of supposing that because in theory the environment is "suitable," say, for the production of a certain crop in a certain area the crop will necessarily be grown there on a large scale or even on a more modest basis. Within any given physical framework the actual use to which the land will be put depends predominantly upon what the economist calls "opportunity costs," which may be defined as the costs of the forgone alternative. In other words, land use and land values depend upon competing uses, and the greater the amount of competition the higher is the value of the land likely to be. Such competing uses may well include possible urban development or mineral production, and when these enter the reckoning land values are apt to rise sharply. But even when we are dealing simply with crop production the number of alternative uses may be high. The value of a piece of land in, say, Sumatra which is suitable for rubber cultivation does not depend solely upon the price of rubber but also upon the rewards to be secured from competing uses—the growing, perhaps, of maize, groundnuts, sugar cane, pepper and coconuts, to name but a few possibilities. In a cash economy these rewards will show themselves in market prices, in profits to landowners and in wages paid to farm workers, and it is obvious that the highest profits and the highest wages are likely to be achieved by producing the crops which give the best financial returns. It is to be noted, however, that it does not follow that all the land in such an instance will be under a single crop, at any rate over a large area. To take one possibility, it will almost certainly not be economic to grow a single cash crop exclusively over any particular area because in that case foodstuffs will have to be imported from elsewhere, and this may be expensive as transport costs are involved in addition to the actual cost of the food.

The value of any given resource produced in the humid tropics, as elsewhere, is therefore very largely dependent upon that of other resources, including in many instances resources from distant extra-tropical lands. This is because most tropical products are in competition with goods produced in temperate lands, and effective ceilings are therefore placed upon their prices. We may examine one or two examples to stress the point, one obvious one being the sugar beet of the temperate zone which competes with the sugar cane of the tropics. It is very difficult to compare relative productivities because of such elements as wage inequality, differing degrees of capitalisation, varying land values, tariffs and quotas, and the greater use made of the residue from sugar beet which can yield fertilisers or fodder and so help other forms of agricultural production. Demand for sugar, however, at present is so high that there may well be room for both forms of production for some time to come.

Another example of a somewhat similar nature is the groundnut, the oil from which comes into direct competition with other vegetable oils of which cotton-seed oil, sunflower oil, soya bean oil and olive oil are perhaps the most important extra-tropical examples, while it also must reckon with the detergents produced by the petro-chemical industry which have very greatly cut back the demand for soap. In the same vein, the demand for tropical cereals such as rice must be affected by production of temperate cereals such as wheat, while maize is widely and most heavily grown in the sub-tropics and the possibilities of developing it as an export crop in the tropics is thereby sharply diminished.

Neither must we forget that it is not by any means simply the existence of natural resources in an area which makes economic production possible. This is abundantly clear in activities like manufacturing and mining but it is true also in agriculture. The success of crop farming frequently depends just as much upon the development of improved strains of seeds and stock, upon irrigation, drainage, transport facilities, and upon the most efficient use of the soil as it does upon the physical environment. In extreme cases the soil itself has even been placed in position by cultivators, in mounds (as in many instances of cassava growing in the humid tropics) or in terraces (as in some of the Rhineland vineyards). This brings us back again to an earlier point, that it is only by the successful marriage of natural and human resources that maximum yields can be secured and successful development take place, and where one or both partners are deficient success may well be impossible to achieve.

THE ASSESSMENT OF NATURAL RESOURCES

There is inevitably in the preparation of a chapter like the present one a temptation at this point to furnish lists of actual resources (particularly mineral resources) and products arranged on a geographical basis as an indication of the presumed wealth or otherwise of the various territories with which the book is concerned. This temptation will be resisted for two main reasons.

1. Such information is available elsewhere in books of economic and regional geography. Simply to abstract and present here this material would be tedious for the reader and unnecessary.

2. Such a list is bound to be to some extent misleading because a mechanical list can give little idea of the real worth of any resources at any given time. This is partly because even the most carefully compiled inventory is out of date even before it can be published, but there are more fundamental reasons which can be illustrated by a few actual examples.

We may take first the example of mineral deposits which in a number of cases have provided very substantial economic benefits for developing territories including many in the humid tropics. It can too easily be assumed that the mere known possession of such deposits is enough to initiate mining activities, yet this is by no means the case. It has long been known, for example, that very large reserves of iron ore exist in Sierra Leone near Tonkolili and in Nigeria near Lokoja, which lies at the confluence of the Niger and Benue rivers. The present writer (Jarrett, 1974b) has examined the Sierra Leone case and it seems that mining is unlikely to commence until world supplies of iron ore become relatively scarcer than they are at present and prices rise. Although the Tonkolili ores are richer than those which have been mined for many years at Lunsar, 130 km to the west, the sheer cost and the difficulties of extending the railway eastwards from Lunsar is prohibitive under present conditions. Similarly, although the Lokoja deposits are less than 560 km from the Gulf of Guinea and not far from the River Niger, transport costs are at present far too high to make the venture of mining them economic.

In some cases a change in technology is essential before an ore body can be exploited. For many years, to take a specific example, the copper ore at Kanshanshi, Zambia, could not be treated by existing methods and the ore therefore remained in the ground. It was not until the development of the new Torco process that it became possible to treat the ore, and mining activites could then be commenced.

Part of our case, then, is clear; the mere listing of known actual or potential natural resources may produce a false picture of present or impending prosperity. But the reverse is also true: an absence of listed resources may give the impression of poverty and suggest little hope for future progress, an outlook which might be quite unjustified. The point may be illustrated from the story of the production of rubber in Brazil and in south-east Asia.

Until the beginning of the present century world demand for rubber was very restricted and it was met from the collecting of the wild product from the Amazonian selvas. During the early 1900s, however, there was a very large increase in demand resulting mainly from the tremendous development of the motor-car industry (Jarrett, 1974b). The story of the establishment of rubber production in south-east Asia, particularly in Malaya, Sumatra and Java is well known (*see also* p. 118 below), some of the output coming from plantations and some from peasant smallholdings, yet it is interesting to note that despite the generally prevailing high prices the output of wild rubber from Amazonia declined—one might say deservedly so in view of the horrific record of unscrupulous exploitation of local populations by masters and merchants. So this apparently favoured South American region, the home of the rubber tree, fell right behind in the production race while other areas which did not appear to enjoy particular advantages drew ahead. Many of the soils used for production, particularly in Malaya and Sumatra, were not especially fertile, while neither country could command the indigenous labour force which is essential for rubber gathering, even on plantations. Soils, however, were coaxed into production and labourers were attracted to the producing areas from southern India, China and Java. But as Bauer and Yamey (1965) observe, no resource survey carried out in the areas mentioned above in, say, the late 1890s, could have held out any hopes that these areas would within a few years become leading world producers of this important tropical product.

If we bear in mind the foregoing points we shall realise that a list of natural resources can be of real value only if carefully compiled notes accompany each entry to set things in perspective—and even then potential resources could hardly be included! The compilation of such a work would itself be a major undertaking and would result in the emergence of a massive treatise very comparable with Chisholm's famous compendium. We shall, therefore, not attempt this task but instead make a few comments upon one or two of the basic resources of the humid tropics, commencing with water.

WATER

We are particularly concerned in this study with the humid tropics, and anyone coming to this topic for the first time may well be forgiven if he assumes that water supply will not be an urgent or a difficult problem, especially as some writers argue that the humid tropics comprise areas in which cultivation can be carried on without irrigation. This may lead us to suppose that water supplies will not be scarce. This assumption is not, however, justified for water supply and water control are among the leading problems of the humid tropics, particularly in the tropical monsoon lands and in the wet-and-dry tropics. The main reasons for this were in fact set out in Chapter I, and they include rainfall variations from year to year, the incidence of a dry season, and very high evaporation rates which are due in part to high temperatures and in part to the effects of drying winds. We should also note that it is quite impossible to be precise about the limits of cultivation without irrigation for the water requirements of different crop plants vary greatly. It is recorded, for instance, that millet can be grown as a rain crop near Lake Chad where there is an annual rainfall of only 356 mm (14 in.), but few if any other crops could successfully be grown under such marginal conditions without irrigation.

Annual and monthly rainfall variations such as those given in the last chapter for Banjul are clearly of vital importance, for the success or failure of the year's harvests very largely depends on them. In the past variations of this kind have in poor years resulted in devastating famines and great loss of life, and even in our own day this danger is by no means entirely absent. The very unequal distribution of rainfall throughout the year also greatly adds to the problem of water supply, and while the problem is naturally more acute where the dry season is long and severe (as in the well-known case of the Sahel region of West Africa) it can exist even in low-latitude (equatorial) lowlands. For instance, Fig. 3 shows us that even in parts of the Amazon lowlands, there is a considerable part of the year when rainfall is below 76 mm (3 in.) a month, a figure regarded by Garnier and others as a critical minimum for vegetation growth. In Manaos itself this "dry" period lasts for about 3 months of the year and the same is true at Belem though no month statistically is completely rainless.

We have earlier (p. 29) commented upon the significant amount of tropical rain which falls with very high intensity. This causes a great deal of loss through rapid run-off, especially on sloping ground, as the soil cannot quickly absorb the masses of falling water, and the value of the rain for crop growing is much reduced. It is particularly unfortunate that the first rains of the wet season upon which soil moisture re-charge depends are frequently notable for their high intensity. Another point is that the effect of a heavy storm is frequently felt only over a limited area, so that while some districts are being almost submerged beneath falling streams of water, adjacent areas can sometimes remain parched and rainless. After such deluges the hot sun quickly makes its effects felt and in the intense insolation much moisture is rapidly lost through evaporation; the balance of water available for agricultural purposes is therefore considerably reduced. It will now be clear that the actual value of rainfall in the humid tropics is frequently a great deal lower than recorded figures may suggest.

Evaporation rates and the effects of drying winds can be very high indeed during the dry season. Farmer (1957) has shown that in Ceylon the period of drought coincides with the hottest and windiest time of the whole year and the resulting evaporation rate is therefore very high indeed; this is, in fact, a very usual experience in the humid tropics. Over most of West Africa, to take another example, the desiccating harmattan blows during the dry season, consistently in the northern parts of the region and more intermittently farther south, and this results in tremendous water losses through evaporation. It is important to note that the tropical dry season is not simply a time of year when it happens not to be raining. Relative humidities fall to very low levels and the affected areas become thoroughly parched as a result of the ensuing high evaporation rate. We have commented in Chapter I upon the extreme discomfort which the inhabitants can suffer as a result of the excessively dry air during this season.

Hodder (1973) draws attention to the probably considerable reserves of untapped groundwater which frequently exist at depth in many tropical areas. It is common (and correct) to associate groundwater reserves with water-bearing sedimentary strata (aquifers) but it is not correct to infer that sedimentary rocks are the only ones capable of retaining groundwater. In the tropics the weathering of basement rocks can proceed to very great depths (depths of up to 30 m—100 ft— have been reported from parts of Sierra Leone) and groundwater reserves frequently occur at depth in the weathered material. The resulting normally stable water tables can be reached by deep wells and these will provide water even during the dry season when all but the largest streams and rivers dry up. Such wells have been sunk in parts of the Gambia where they have abundantly proved their value, and there is no doubt that much greater use could be made of groundwater than has been the case in the past.

In connection with the third Five-Year Plan for India it was estimated that an annual rainfall equivalent to roughly 370 million ha m is experienced over the country as a whole. Of this total, about 123 million ha m are rapidly and irretrievably lost through evaporation, 167 million flow into streams and rivers, 43·2 million are retained in the surface soils, and a further 43·2 million sink into the ground to replenish underground reserves. Much of this water cannot, it is true, be used, particularly that lost through evaporation and much of the river water, but it is estimated that only about 36 per cent of the available surface water is at present being utilised together with rather less than 20 per cent of the annual replenishment of groundwater supplies. While these are very approximate figures which take no account of local conditions there can be little doubt of the message which they carry.

It is perhaps relevant to note that in some parts of the humid tropics the problem consists of knowing how to cope with the surplus water which from time to time causes havoc and devastation. Bangladesh provides an outstanding case in point. In other areas again the important question is how to control the movement of water and so to make it available for use. In some measure this may be achieved by growing crops best suited to local conditions, for water requirements of different tropical crops vary widely, but on a broader canvas some form of direct water control is often desirable.

In many cases some measure of water control can be achieved by various traditional methods such as the simple expedient of damming small streams and rivers. Brushwood dams are frequently constructed across small rivers in south-eastern Asia, for example, and the impounded water is used for rice growing. In peninsular India and Ceylon the tank is a common feature (Fig. 8); this is an earthen embankment thrown across the line of a stream or damming the outlet of a natural depression. Different kinds of *shaduf* are in use in many parts of the Asian and African tropics. Many writers have stressed the importance of all small-scale works which can be built, maintained and operated by local people as opposed to the large-scale enterprise which requires vast amounts of capital and skilled labour, both of which are scarce commodities in most tropical territories.

The large-scale enterprise, however, has come to stay and recent schemes have been mainly of the multi-purpose type. One of the earliest of these was the Volta Scheme of southern Ghana which not only produces hydro-electricity but also encourages food production through the provision of water for irrigation and also has made possible greatly improved communications, particularly in a north–south direction over the waters of Lake Volta. The Damodar Valley Scheme in

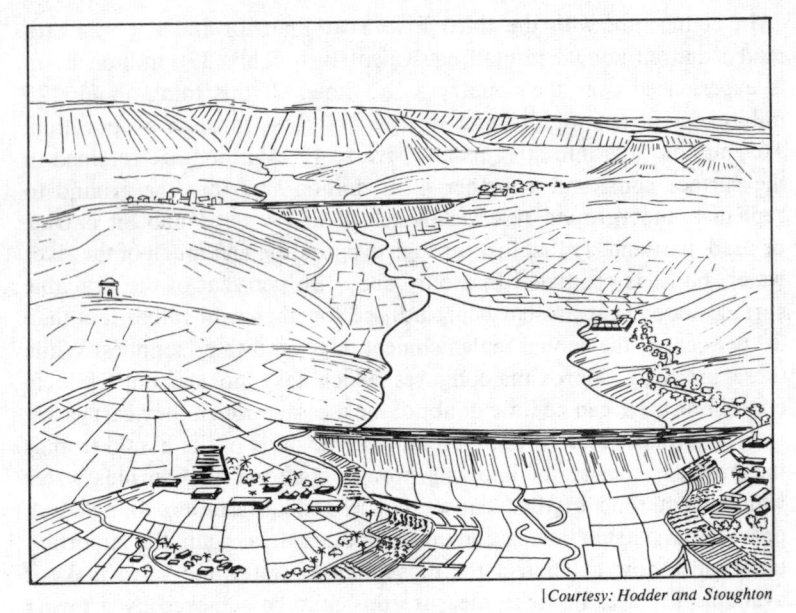

FIG. 8.—A typical rural scene in The Deccan.

[Courtesy: Hodder and Stoughton]

India is in some respects comparable, an important feature being that the various reservoirs which form integral parts of the scheme help to absorb sudden surges of flood water during the rains while the river is no longer dry during the dry season. This is a characteristic feature of many multi-purpose schemes and it often greatly improves the river from the point of view of navigation while, at the same time, averting the former dangers of flooding with consequent crop losses and destruction.

Other comparable schemes are in operation in many parts of the humid tropics, including tropical Africa and in parts of South America, particularly in Brazil and Venezuela.

SOILS

The soil covering of the earth's outer shell may be defined as that part of the crust which extends downwards from the surface as deeply as living organisms penetrate. Soil is not simply a static, unchanging mineral layer. It is dynamic as some of the soil material is constantly being removed (mainly by denudation) while other material is constantly being added because of weathering of the underlying rock, because of the deposition of particles brought from elsewhere, and because additional accretions accrue from the fall and decay of vegetation. Chemical reactions occur within the soil and these are speeded up by increasing temperatures; a rise of about 10° C in surface temperatures will result in a doubling in the rate of chemical reactions at normal temperatures, and it is not surprising, therefore, that chemical changes within the soil and during rock weathering are frequently three times as rapid in the humid tropics as they are in cool climates. A further point to bear in mind is that landscapes in the humid tropics are frequently of very great age, for example over the plateau surfaces of Brazil, the continent of Africa and the Deccan of India, so that weathering and soil formation have in such cases reached a very advanced stage. There have been no interruptions to the process of soil development such as there have been in cooler areas with the onset of glacial periods which have modified soils and soil processes generally.

It is now fairly widely recognised that tropical soils are for the most part inherently infertile and it is argued that this infertility forms a very real limit to agricultural development in tropical areas. In general terms the reason for this is clear and has previously been referred to—the high rate of weathering and chemical change under the prevailing warm, humid conditions. This encourages the leaching of bases and of silica, leaving a layer rich in iron and aluminium compounds at or near the surface; there are, in fact, two main types of mature soils which form under these conditions: kaolinite, formed over acid rocks, and ferrallite, formed over basic rocks.

Unfortunately, tropical soils are normally weak in their adsorbtive capacities. To point a contrast we may note that clays of temperate latitudes are montmorillonites, hydrated silicates of alumina which allow the penetration of water and which attract such valuable elements as potash, ammonia, lime and magnesia. These adsorbtive clays form the basis of many fertile soils. Tropical soils, however, such as the kaolinites and ferrallites have lost more of their silica, they cannot easily absorb water, and they do not easily attract and retain the fertilising elements such as those mentioned above. Tropical soils contain a high proportion of kaolinite, quartz grains and hydroxides of iron and aluminium, and these inert fractions allow penetrating water easily to remove in solution any fertilising elements and humus which it may initially carry. Such soils therefore quickly deteriorate if they are cultivated.

Many tropical soils have a strongly reddish or yellowish coloration imparted by various iron compounds. Such soils are widespread on freely-drained sites, both in rain forest and in the savanna zones, and it is these which form the broad latosol group. All latosols are freely-draining, have an acid reaction, are moderately to highly weathered and are moderately to highly leached (Young, 1974, 235). Textures vary widely from loamy sand to clay depending upon the parent material, and agricultural potential varies between high and very low. It is thus not possible to generalise about this broad soil group. We might, however, note that in latosols the iron remains dispersed throughout the soil profile, but if drainage is impeded the development of a strongly-enriched iron horizon is likely to ensue.

If, under these circumstances, iron and aluminium oxides are concentrated into a definite layer a lateritic horizon develops, and it has traditionally been agreed that if this horizon is exposed by normal weathering or by other means (perhaps as a result of accelerated soil erosion consequent upon cultivation) it will invariably harden into a sterile crust often known as *carapace latéritique* or lateritic crust, though this view has recently been challenged. Of the existence of these crusts, however, there is no doubt, for they are well-known in many parts of West Africa where they are sometimes known as *bowal* and in Bengal where the name *khoei* is given to them. They are impermeable and they weather only very slowly to give a patchy and very thin soil which supports only a degraded scrub type of vegetation. It is feared by many observers that such areas are being extended by ill-regulated agricultural activities.

Gourou (1968) demonstrates that chemical analysis indicates the mineral deficiencies of tropical soils, and he quotes Baeyens who asserts that if much of the land occupied by profitable plantations in Zaïre could be transported and set down in, say, Belgium, it would support only heath and utterly barren moorland. He also shows that tropical rivers are typically poor in both dissolved and suspended matter because of the exhausted state of the soils in the areas over which they flow. Thus, the Amazon carries 0·1966 g of suspended matter per litre and the Zaïre 0·0385; these amounts may be contrasted with those measured on the Mississippi (0·638 g), the Yangtse Kiang (0·930 g) and the Hwang Ho (4·8 g). An important corollary of this is that rivers in the humid tropics cannot build extensive and rich alluvial plains as they can in temperate regions. The so-called Amazon lowlands, for example, are not rich and they have not the true form of an alluvial lowland, while the area known as the Inland Delta of the Niger between Segou and Timbuktu has proved very disappointing as a crop-producing area, even with the help of irrigation, owing to the poor quality of its soils.

It would, however, as we have already seen, be incorrect to assume *(a)* Relation between soils and relief.

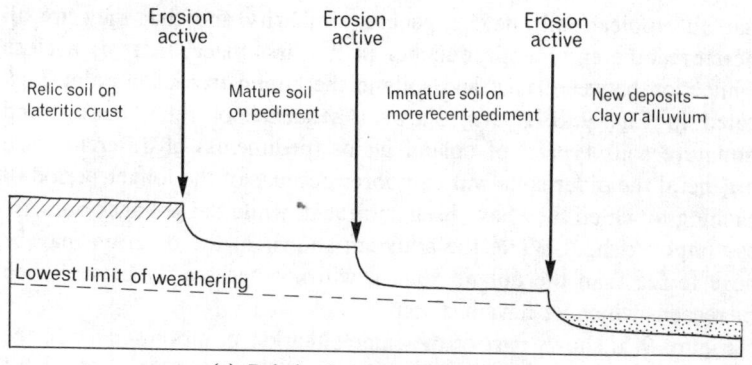

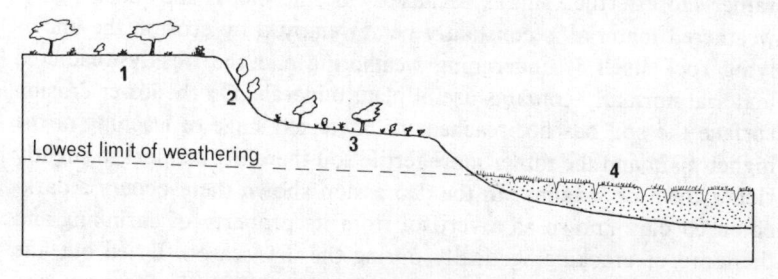

Labels: Erosion active; Erosion active; Erosion active; Relic soil on lateritic crust; Mature soil on pediment; Immature soil on more recent pediment; New deposits—clay or alluvium; Lowest limit of weathering.

(b)
1. Ferrallitic soil in deeply-weathered material supporting savanna.
2. More fertile soil formed under native erosion supporting forest.
3. Lower-level immature soil supporting richer savanna than in 1.
4. Vertisol.

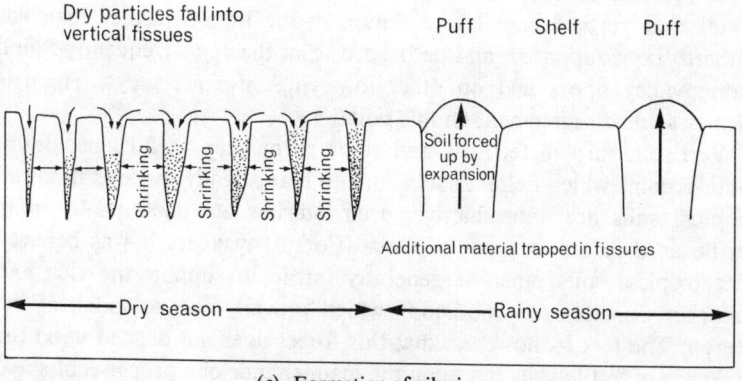

(c) Formation of gilgai.

Labels: Dry particles fall into vertical fissures; Puff; Shelf; Puff; Shrinking; Soil forced up by expansion; Additional material trapped in fissures; Dry season; Rainy season.

Fig. 9.—Relief and soils in the humid tropics.

that all tropical soils have a general similarity and that they are all therefore inherently unproductive. In the first place, there is a close connection between relief and soils in the humid tropics, a point illustrated in Fig. 9. Fig. 9(a) shows a sequence of relic, mature and immature soils typical of upland plains (pediments) of different ages; in general the older soils will be poorer because of the longer period of leaching to which they have been subjected, while the younger ones are less impoverished. While the alluvium shown in the diagram may be more fertile than the upland soils it will not necessarily be very rich for reasons which we have just seen.

Figure 9(b) shows part of the same situation in more detail. On the highest upland plain worn out ferrallitic soil is widespread but on the scarp slopes which are undergoing comparatively rapid erosion soils are rather more fertile. This is because while the oldest and most highly-weathered material is constantly being removed by erosion the underlying rock itself is undergoing weathering and the freshly-weathered material normally contains useful plant minerals. On the lower erosion surface the soil has not reached the advanced stage of leaching of the higher plain and the rather more fertile soil therefore supports a slightly richer form of savanna. In the depression shown there occurs a dark-coloured clay known as a vertisol from its property of shrinking and therefore of cracking vertically during the dry season. Dried out fine particles are blown into the vertical fissures and partly fill them, and during subsequent rainy seasons soil expansion therefore meets with resistance as the vertical fissures are partly filled, and as a result low soil ridges or knolls known as "puffs" are forced upwards. Between them lie lower and flatter sections known as "shelves" (Fig. 9(c). The ensuing form of micro-relief is known as *gilgai*. Young (1974, 237) states that vertisols are less common in the humid tropics than has hitherto been supposed, and he believes that the clays frequently found along valley floors and on other low-lying sites are gleys. There is clearly some disagreement on this point.

Vertisols vary in fertility, and some form very useful agricultural land, a point which helps to substantiate that already made that not all tropical soils are unproductive; even laterites are occasionally quite fertile as they are in parts of Cuba. For many years it was believed that tropical soils must be generally fertile to support the rich and luxuriant forests and woodlands which prevail over so much of the region. The fact is, however, that this forest does not depend upon the presence of fertile soils but upon the maintenance of a proper ecological balance. This fact is now well known and need not be elaborated here, but it depends upon the production of vast quantities of decaying and decayed organic matter by the forest itself which is therefore able to maintain an admittedly rather low level of humus in the surface soil layers. The comparatively low humus level is due to the rapid decomposition which prevails in the warm, humid conditions, but as long as the cycle of humus, tree growth, waste vegetation and humus is unbroken the forest flourishes as an integral part of this rather fragile biological balance. Mohr believes that beneath rain forest equilibrium is established between the production of humus and its destruction when the soil temperature is about 25° C (77° F); under cooler conditions the amount of humus in the soil will increase and under warmer conditions it will decrease, and it is fortunate that tropical soils beneath a rain forest cover do in fact maintain a temperature very close to Mohr's optimum figure. As long as this balance is maintained, therefore, the lush forest is perpetuated, but on the other hand if the forest is destroyed the soils will rapidly deteriorate.

This, then, is the great lesson to be learned, that tropical soils often can be very productive if they are used in the most effective manner and if the correct ecological balance is struck. It now seems likely that "temperate" conceptions of soil fertility may not be applicable to the tropics though fertility does, of course, vary in these regions as it does elsewhere. On the whole it is probably true to say that mature upland soils are inherently poor while immature soils, upland and lowland, are of better quality as they have suffered less from leaching. South-eastern Asia is unusually well favoured in this respect for the large rivers, notably the Red River, the Mekong, Menam, Irrawaddy, Brahmaputra, Ganges and the Indus which flow from the comparatively recent fold mountains of Alpine age carry substantial amounts of fertilising silt, and they have therefore built up fertile plains and deltas over which soils are immature. There is nothing comparable in tropical Africa or South America. In addition there are the fertile volcanic soils of Java which are still being nourished by continuing vulcanism, while the *roxas* of São Paulo, derived from basaltic lavas, are also very productive. Even so, the importance of maintaining the most efficient biological balance in the tropics can hardly be overstressed, for as Moss (1963) has pointed out the bio-geographical balance in the humid tropics is frequently a most delicate and fragile one, and an apparently minor disturbance can bring about the collapse of the whole system.

By way of summary we might justifiably argue that it is very difficult at present clearly to assess the potential of tropical soils as we await the results of more research. On the one hand Gourou (1968) avers that only a small part of the tropical lands is naturally very fertile, while on the other hand Hodder (1973) states that there is no evidence for postulating inferior fertility in tropical soils. This may possibly be, but there is little doubt that in our present state of knowledge and experience agricultural productivity is far more difficult to attain in the tropics than in temperate lands. Even so, we must admit that instances of high and regular yields under modern productive methods in the tropics are not lacking, for example the production of sugar cane in Hawaii, palm oil in Sumatra and rubber in Malaysia.

Soil erosion

The menace of soil erosion is a very real one in the humid tropics, water erosion being more important than that due to wind. It is of course a simple fact that surface erosion has proceeded throughout geological time for without it there could be none of the normal processes of denudation. What we are here concerned with, however, is the vastly accelerated erosion induced by man's interference with the biological balance of nature. As the natural vegetation cover suffers interference, and perhaps is even temporarily destroyed for purposes of cultivation, both sheet and gully erosion rapidly develop under humid conditions. Sheet erosion takes place when a sheet of water moves down a slope—a not unusual occurrence in a region subject to high intensity rainfall—and removes particles from the topsoil. It is harmful because it is the finer and nutritionally more valuable particles which are removed first so the soil affected becomes steadily coarser, less retentive of nutrients, and less fertile. As water flow becomes localised gully erosion develops, and serious gully erosion can ruin an area as far as cultivation is concerned. Sometimes gullies are of impressive size with tributaries running into major gashes, the whole forming an intricate dendritic pattern. Deep gullies may destroy both soil and subsoil and they may unfortunately lower the water table. On the whole it is possible that the more insidious sheet erosion is on balance the more menacing because it can continue unnoticed for a long period, whereas the effects of gully erosion are rapidly obvious.

The golden rule to be observed is for farmers to retain a vegetative cover over the soil though this is not always possible under cultivation. Inter-tilled crops such as maize, tobacco and cassava which are widely grown in the humid tropics consist of separately standing plants which permit the free movement of water around them; it is therefore desirable to rotate such crops with cover crops which form a more continuous carpet over the soil. Useful cover crops include grass, leguminous crops, sugar cane, sweet potatoes and pineapples, and these all inhibit the movement of surface water. Tree crops also provide good soil protection when they are well established but care must be taken when the trees are young and before branch and leaf cover are well developed. Clean weeding is rarely if ever a good practice in the tropics. Particular care is needed when cultivating sloping ground, and contour farming or terracing is sometimes desirable though this becomes difficult on slopes steeper than about 15 degrees.

It is not always realised that stock rearing can induce soil erosion if it is not carefully controlled for overstocking inevitably produces overgrazing and the consequent destruction of the vegetative cover. Care must be taken, too, when moving stock on the hoof, for it is clear from experience in southern Africa, to take one example, that if animals are driven each night to the protection of compounds and released each evening their habit of keeping to well-trodden paths will initiate the establishment of large gullies as storm water surges along these tracks from which the grass covering has been trodden away. And erosion which begins in the compound itself and along these tracks will usually spread more widely. The clear solution to this problem is, of course, to avoid overstocking but reduction of stock numbers is not easy in communities in which the mere possession of animals confers prestige and wealth, and in which cattle are kept therefore in numbers as large as possible rather than as integral parts in a controlled grazing or mixed farming economy. We move at this point sharply away from the physical basis into the realms of sociological and cultural customs.

We have said enough to emphasise the tremendous importance of soil erosion in the humid tropics, and we should bear in mind that the problems which it poses may well become more urgent as populations increase, as more land is brought under cultivation, and as more of the protective vegetation cover suffers from human interference.

VEGETATION

The main types of vegetation met with in the humid tropics have already been described and we need not add to what has already been said. We have noted that while it is sometimes argued that tropical rain forests form a barrier to human progress there is almost certainly less force in the argument than is often supposed. In other words, the "negative" significance of tropical vegetation is probably not very high except, perhaps, in cases such as extensive swamps.

With regard to the natural vegetation of the humid tropics, it is true to say that the present-day economic value of tropical forests is not very great. The tropics produce only about 10 per cent of the world's supply of timber a figure which is extremely surprising when we remember that it is estimated that the productive tropical forests cover an area of some 11·9 million km² as opposed to the 14·0 million km² of the productive temperate forests, while on average the tropical forest produces 12·5 m³ of timber per hectare as opposed to the 2·8 m³ of the temperate forests. There are, of course, reasons for this comparatively poor performance. Hardwoods are dominant in the tropical forests and these are very difficult to cut and trim, while at the same time the greater part of world demand is for soft woods largely because the big markets and the large-scale manufacturing regions where most timber is used are located in the temperate (soft wood) belts and the timber-using technologies are therefore not unexpectedly geared to the utilisation of temperate soft woods; this inevitably puts the humid tropics at a severe disadvantage. Attempts to exploit a single species of tree meet with difficulty for pure stands are exceptional in the rain forests (p. 19 above). Finally, only limited parts of the rain forest lands have the lines of communication necessary to move such a heavy and bulky commodity as timber. Some writers have recognised concentric zones of progressive exploitation in rain forest areas, especially in the highly-populated parts of India, in the Indochinese peninsula, and in Malaysia, while in Africa it is seaboard states with useful rivers which form the chief timber-producing regions. Examples include south-western Ghana and Gabon where the new trans-Gabon railway is expected to open up extensive areas of rain forest in the interior and to increase the output of timber (p. 195 below).

Other forest resources include wood for fuel, cellulose, resins and gums, camphor and others of less importance. As obstacles to the exploitation of the rain forests are removed and as the economic incentive to develop them increases we may well see them become areas of considerable importance in global economic affairs, but great care will be needed, for example, not to overcut, for as we have already seen disastrous results can accrue from widespread forest destruction in the humid tropics. Forest conservation and re-afforestation schemes are becoming more essential every year for if once the rain forest disappears there is no guarantee that it will regenerate itself—indeed, it certainly will not during the spans of several human lifetimes (p. 19 above).

Specific types of economically useful rain forest trees can, of course, be introduced into a region where they do not naturally occur and much economic development within the humid tropics has been made possible from the practical consequences of this. Plantations, which can almost be viewed as artificial forests, have been widely established and these include rubber plantations in Brazil, south-east Asia and (to a lesser extent) Liberia and Nigeria; palm oil plantations in Zaïre and southeastern Asia; and cocoa in West Africa (mainly in Ghana, Nigeria, the Ivory Coast and the Cameroun Republic). Plantation crops which are based on bushes rather than on trees include tea, cotton and coffee. There is little doubt that plantation agriculture could be considerably extended in the humid tropics but it is not altogether clear that such an extension would be welcome on other grounds. The main points at issue are discussed in Chapter V.

Another valuable forest tree which has been transplanted elsewhere with great success is the Asiatic teak with its straight, branchless trunk and its durable timber. Much natural teak forest, as in northern Thailand, has been destroyed because of burning for shifting cultivation but such forests are now being preserved. Teak is now planted in parts of West Africa (in southern Nigeria, for example) and is proving a very successful "crop."

The question whether it will in the future be economically advantageous to practice sylviculture as has been the case for centuries in western Europe is becoming an increasingly pressing one. Few, if any, of the forests of western Europe are natural and tropical forests, too, could be so managed that they would produce large and dependable amounts of the kinds of timber needed for economic purposes—for constructional and domestic purposes, for veneers and for paper manufacture, to name but a few examples. A network of roads could be constructed in these managed forests to facilitate the movement of timber, and it is estimated that with care a normal mixed rain forest could be transformed into a commercially viable production unit in less than a century.

An excellent example of the need for sylviculture and of the possibilities which might ensue comes from Brazil, a land of natural forests on the grand scale. Even forests of this magnitude, however, are not limitless and there are many observers who fear that the present scale of forest destruction in the north of the country to open up land for cattle rearing has a very dangerous element about it. This development is discussed later in Chapter V.

What is not open to doubt is the decimation of forests in the state of Minas Gerais to provide charcoal for the iron and steel industry. Demand is for 15 m^3 of charcoal annually, and this is expected to rise to 24 m^3 by 1980—and six trees are required to produce a single m^3! So widespread has the resulting devastation been that supplies now have to be despatched to the iron and steel centres by road for distances up to 500 km.

A massive planting programme has now been begun and the central

Government, the state Government and the iron and steel companies are all participating in this. Natural conditions are favourable, for thanks to the climate, altitude and soil trees grow at surprisingly rapid rates; research and experimental plantings have shown, for instance, that eucalyptus trees will grow to maturity in as little as 7 years, and pines in between 9 and 12 years. It is expected that eventually there will be sufficient timber produced not only to meet the demands of the iron and steel industry but also to supply four or five cellulose and paper factories which should provide substantial surpluses for export. One and a half million hectares will have been planted by 1980, much of this consisting of land not suitable for agriculture, and employment may ultimately be provided for some one million workers.

MINERALS AND POWER

In theory, mineral and power resources should be as well distributed in the humid tropics as elsewhere in the world, yet until fairly recently tropical lands did not produce either minerals or power in amounts in any way commensurate with their size. Even today production remains strongly localised.

The reason for this situation is not difficult to see. It stems from the fact that the demand for minerals and power is derived overwhelmingly from industry, and modern industrial development began in the temperate parts of the world. Even today the major industrial regions are all located in temperate lands and we see only the beginnings, promising as these may be, of comparable developments in the tropical world. In view of this overall geographical pattern it is only to be expected that industry sought, and until fairly recently usually found, the necessary resources of raw materials and power in the temperate regions themselves. In earlier years sufficient quantities of these essential requirements were available within the confines of the various industrial states and it is only in more recent times, particularly since the close of the Second World War that supplies in the older producing areas have become totally inadequate to satisfy the enormously increasing demand. Supplies have therefore been sought further afield, even as far as many countries of the humid tropics.

This tendency towards ever-increasing growth strengthened as the industrial nations became more affluent, for it is inevitable that as this happens the market for the products of industry expands. There is thus great pressure to expand production and this in turn leads to greater demands, among other things, for raw materials. In the face of this expansion local resources frequently prove inadequate and the search for new sources of supply is prosecuted with ever-increasing vigour. This is where hitherto underdeveloped territories are brought into the industrial framework as their mineral resources are eagerly sought after.

But the story does not end there for the new mining interests begin to play an important part in stimulating economic expansion in the L.D.C.s themselves, at first on a very modest scale and later more vigorously though such expansion and development is never uniform over any territory as a whole. It occurs in the "pockets" or "islands" of development which we noticed earlier (p. 7 above), and which are set in the early stages in a matrix of general underdevelopment. Such pockets commonly develop in mining areas though they may establish themselves as a result of intensive agricultural development as crops are grown for export, or they may stem from some other advantage such as that of being associated with a port or other urban centres. Such pockets normally attract capital, labour, enterprise and infrastructural development in what Myrdal (1957) has called a backwash effect, and they customarily, at least for a time, stand starkly out as isolated islands of economic and social advancement. Hance (1964) has pointed out that there is in the case of Africa great unevenness in the size and distribution of these islands which are commonly separated from each other completely or at best linked by the most tenuous strands. In time, however, a "spread effect" will come into operation as the capacity of the pockets of development to provide goods and services for local consumption is outstripped by the expansion of the domestic market, and development will then begin to open up in other areas. Slowly, therefore, economic and social advance spread to other areas within the territorial framework.

The tracing of this sequence of events has taken us a short distance from our present theme but, sketchy and incomplete as it is, it helps to explain the present pattern of development in many territories in the humid tropics with regard to mineral and power production. It helps to explain why, in the comparatively affluent post-war world, many of the L.D.C.s have gained very considerable importance as suppliers of minerals and it explains the increasing concern now being shown in those same territories to increase power supplies as increasing economic development stimulates demand. We may expect these trends to continue except in the perhaps unlikely event of a major world economic recession of lengthy duration.

One word of caution is, however, necessary. The finding of new mineral deposits and the subsequent opening up of new mines involves a very risky series of undertakings. In the first place a very great deal of money must normally be spent in following up promising discoveries of minerals in order to decide whether a potentially profitable mine exists. It is then necessary to estimate the net cash flow that the mine can be expected to induce over a period of years. At first this will be negative as capital is put into the new enterprise, and later, as profits begin to accrue, allowance must be made for taxes, replacement capital, capital repayments and interest on capital. In addition it must be realised that a time will come finally when all the workable mineral has been extracted and when the mine will be worthless, despite all the investment which has been put into it, and an allowance must therefore be made for the progressive depreciation of the whole mine, including its equipment.

We have clearly moved far from the discovery of the original ore body for these matters involve political and fiscal as well as purely economic considerations. Political questions which must be considered by any potential investment group include the stability of the territorial government concerned, its past record and its probable future policy for dealing with foreign investors. The fiscal problems may be enormous for governments in these days exhibit voracious appetites for taxes and this has led to the growth of a bewilderingly complicated network of international treaties designed for the avoidance of double taxation. The mining company executive must evaluate the efficacy of this network in any particular case, and before any decision on the commencing of mining can be taken a very great deal of corporate skill must be devoted to considering legitimate ways of avoiding "tax tripwires" (*The Times*, 16th October 1970) because success or failure in this exercise can determine the success or failure of the whole enterprise.

Economic considerations, of course, are also heavily involved in the various costing devices designed to forecast the flow of money which the mine can be expected to generate (the net cash flow referred to earlier), while the securing of sufficient capital to finance the project, at least in the early stages when the net cash flow is negative, can be crucial. There are times when after considering all the hazards involved in mining enterprise today one marvels that any such developments take place at all!

It is not the purpose of this section to present a gazetteer-like list of minerals produced in the various countries of the humid tropics, nor of the energy resources of these lands, but one or two specific examples may be given by way of illustration. Perhaps the basic industrial raw material is iron ore and the humid tropics have very substantial reserves of this. Large amounts are exported by Venezuela, Liberia, Sierra Leone, Angola and Brazil, while Brazil and India also produce large quantities for use in their domestic iron and steel industries. Other territories such as Nigeria also possess large reserves.

Another key metal of today is copper, for which there is at present no substitute in the electricity industry and for many other purposes. Central Africa is easily the world's largest supplier of this metal, the leading producers which share what is physically the same copper belt being the Shaba (formerly the Katanga) province of Zaïre and Zambia; significant amounts are also produced in Rhodesia. It is an ironic fact that the world's largest known single copper deposit is located far from the open sea near the centre of the world's second largest continent, and it is not surprising that political troubles have from time to time hindered production (Jarrett, 1974a). Bauxite is another ore of which the humid tropics have large reserves, notably in Guyana, Surinam, Jamaica, the Republic of Guinea and elsewhere, while tropical territories, especially Malaysia, Indonesia, Bolivia and Nigeria, supply the greater part of the world's tin. These are by no means the only mineral deposits in the humid tropics but they will serve to show the very great importance of these lands in mineral production.

One hitherto generally neglected point in connection with mining which we have so far simply mentioned but which may become increasingly important is the status of any given mining area after the mines have closed and the mineral supply exhausted. The former mines are then completely valueless despite the possibly heavy capital investment which they attracted during their working life. So far, largely because land is frequently not a scarce commodity in the humid tropics, this aspect of the matter has generally been neglected but with rapidly increasing populations and an increasing urgency regarding food supplies it may become vital. This has been the case, for instance, in Britain where today open cast mineral workings must by law be restored so that other forms of activity, possibly farming, may be resumed, and in a country like Sierra Leone where diamond diggings have become very extensive along potentially productive river valleys such land restoration may become of vital importance.

The situation on Ocean Island, more properly known as Banaba, offers a notable case in point. Banaba, which has a total area of only 607 ha (1500 acres), lies very slightly south of the equator in the western Pacific and technically forms part of the Gilbert and Ellice Islands. In 1900 phosphate deposits were discovered there; mining began in 1902 and in 1920 the mining rights were sold for £3·5 million to the British Phosphate Commission, a consortium of the British, Australian and New Zealand governments. A royalties fund was established for the benefit of the islanders who, however, were dispersed by the Japanese invasion of 1942, and after the war they were settled on Rambi, an island in the Fijian group; they are now, however, desirous of returning to Banaba. They do not wish, moreover, to remain grouped politically with the Gilbert Islands which are to become politically separate from Ellice Island, while they resent the fact that much of the money derived from the taxation of their income from the phosphate mining has been used for the benefit of Gilbert and Ellice and little for Banaba. In addition they claim that their royalties have been lower than they should have been because the British Phosphate Commission have sold phosphate at less than its market value and that the Commissioners are not keeping an undertaking to restore worked-out land.

It is expected that the phosphate on Banaba will be exhausted by 1980 but unless action is quickly taken to honour the undertaking and to develop other forms of economic activity the returning inhabitants will be returning to an island which has been so badly used that it will be virtually uninhabitable. One contemporary description of the surface after the cessation of mining activities is that it looks like a badly decaying tooth from which the filling has fallen. Considerable controversy rages over this case and it is not easy to form an opinion at present; for instance, the waters around Banaba are rich in edible fish and this source of wealth will remain though the islanders argue that a new fishing industry should be established by the Commission. It is also true that the phosphate mining has made the Banabans very wealthy, but the final cost to them may be very high.

It now seems that the humid tropics possess greater quantities of the mineral fuels than was once believed though reserves as far as is at present known do not match reserves in other parts of the world. Only India produces large amounts of bituminous coal though production does take place elsewhere, notably in Brazil, Rhodesia and Nigeria. It is now thought that tropical territories may possess about one-third of all known reserves of mineral oil.

The possibilities for the development of hydro-electricity, however, are enormous thanks largely to the nature of the relief and climate. The climatic contribution, of course, lies in the rain though as the rainy season decreases in length with increasing distance from the equator this becomes a less useful asset. The relief is very important. The ancient pre-Cambrian massifs of Brazil, Africa and India (the Deccan) provide numerous possible sites for dams and generating stations, particularly along their edges where rivers cascade downwards towards the sea, or inland where rivers storm their way through gorges or fall from one pediment level to another. In Venezuela, Brazil and India large hydro-electric schemes are now in operation but perhaps it is in Africa that the largest potential lies. Earlier schemes such as those at Kariba (Zambia–Rhodesia), Owen Falls (Uganda) and Akosombo (Ghana) caught the public attention as they were very spectacular pioneer projects but they are greatly eclipsed in magnitude by Cabora Bassa in Mozambique and even more so by Inga, located where the Zaïre river, on its passage through the Crystal Mountains (a passage which has been described by Batchelor (1975) as "among the most daunting stretches of white water in the world") falls 100 m in a 25-km "awesome span of sustained violence." Power will be exported from Cabora Bassa to the Republic of South Africa while even in its first phase Inga is capable of supplying all the present needs of Zaïre from Kinshasha westwards (Jarrett, 1974a). In due course power will be transmitted eastwards to the Shaba Copper Belt.

Such schemes are attractive to the territories of the humid tropics because the unit costs of producing power are so much less than those incurred in coal-fired and diesel power stations; the ratio is said to be 1·2, 3 and 25 respectively. There is also the point that such schemes can be (and increasingly are) made multi-purpose (*see* p. 41 above), but it must be remembered that they raise very real problems, particularly with regard to the securing of the large amounts of capital required.

Chapter III

Human Resources of the Humid Tropics

PRELIMINARY CONSIDERATIONS

THERE are probably few parts of the world which have given rise to so many misconceptions as the humid tropics. This is perhaps inevitable and it goes back to the time which finished, we should remember, not so very long ago, when comparatively few white people worked in these areas and when there were few opportunities for accurate and scientific study. One widespread misconception, for instance, is that tropical areas are very lightly populated, but we know now that this is by no means always the case; we have examined this point in Chapter I and we need not further elaborate it here. It is also quite widely believed that the hot, moist, enervating climates so typical of large parts of the tropics are far from ideal from a human point of view and that they nurture inhabitants who are not capable of sustained effort and who prefer leisure to the performance of any work above the essential minimum—who are, in short, lazy and shiftless. The difficulty is that while many of the inhabitants of the humid tropics certainly fit this description, as do many inhabitants of the temperate lands, it is not a simple matter to identify the cause. The climate may well play its part, but so may heredity, tradition, and the facts of the economic environment in which these people live, and many writers have argued that if human skills can be sharpened and economic patterns developed in which effort receives its due reward, then the people of the humid tropics will respond as others have responded elsewhere. The question remains: how are these aims to be realised?

THE PAST AND THE PRESENT

Any human society is to some extent the child, some might say the prisoner, of its past, and this is particularly true of traditional societies. Old customs and traditions are frequently still observed though the social milieu which gave rise to them in the past may no longer exist. It might be useful to examine some instances of this in practice, noting at the same time some of the results of certain fossilised social customs.

Our first example is that of the extended family which is a social institution widely met with in the humid tropics, and in traditional societies it has much to commend it as a kind of insurance against hard times. Property and income are not viewed as personal possessions but as resources of the whole extended family, and quite distant relatives may be accounted family members. Earners therefore must accept responsibilities towards a considerable number of people, while individual family members who fall on hard times are cushioned by the family as a whole. The converse of this is that an extended family will frequently act together to sponsor a promising youngster and to finance his progress through school and perhaps university, confident that when he begins to earn a good salary all will share in his enhanced income.

There is a great deal to be said for such a system in a traditional society when no public provision can be made for the relief of the needy, when few, if any, scholarships or grants are available for intending students, and when there are no facilities for storing present wealth in anticipation of future needs. Even farm surpluses cannot be stored for long periods and in truly traditional societies they cannot be marketed and the money put by. It is under these circumstances no more than common sense to share out foodstuffs and to use them while they remain wholesome, leaving the future to take care of itself. At a later stage of economic and social development, however, when social services take root and when a cash economy develops, such a system can act as a dead weight upon progress, and it may stifle initiative and enterprise among the young when they know that any fruits of their labours must be widely distributed among family members.

Another reason why the system discourages enterprise among the young is that it encourages, perhaps even demands, that they adhere unthinkingly to the practices of their fathers. When the question of attitudes to the use of fertilisers in an Indian village was debated, for instance, most farmers said that they would not accept such a new practice if their fathers were opposed to it. In eastern Zambia, where ownership of cattle is vested in a kinship group, older group members block attempts to improve grazing and the quality of the animals, and they also refuse to allow any increase in the numbers of cattle slaughtered in order to reduce the overlarge herds; these excessively large herds form too heavy a burden on the available pasture which is thus overgrazed and degraded. On the other hand, the extension of cocoa growing in southern Ghana was greatly helped by the system, as extended families frequently formed units of enterprise with the senior male of the group acting as financier and director.

Bauer and Yamey (1965, 71) point to the importance of various restrictive practices which are of widespread occurrence in the humid tropics. Examples include restrictions upon the movement of villagers and upon the acquisition and employment of skills useful to the community. There is a loose parallel here with the powers of the craft gilds of Europe in the Middle Ages which gained tremendous power through their control over the entry of apprentices to particular crafts and over the professional conduct of their members. Sometimes these restrictionist attitudes lead to very marked divisions within a society, as in the outstanding case of the caste system of India. While it is true that such attitudes typically reflect such specialisms as exist within a society it is also true that when they become too rigid they become repressive and destructive of effort. It is also frequently the case that traditional societies are markedly xenophobic and it is not unusual for open hostility to prevail between different village or tribal groups. In such cases larger-scale specialisms are inhibited because any form of economic co-operation between the different groups concerned is impossible, largely because it is believed to be harmful to the cohesion and the economic strength of particular societies.

Customs and traditions like those just described are deeply rooted and it takes a very long period of painful experiment to eradicate them, as the experience of the United Kingdom and of any developed territory shows. Normally, however, they retreat, albeit slowly and painfully, before the march of economic progress and in theory there should supervene a period of minimum restraint as the old sanctions are cast off and the society moves towards a period of free competition and a market economy. This was the case in Britain and it is fair to say that economic progress was greatly encouraged by the general lack of legal and social restraint at the time of the agricultural and industrial revolutions, unfortunate as were many of the by-products of this *laissez-faire* environment.

This is not, however, the situation prevailing today, even in the L.D.C., whose governments frequently engage in various forms of restrictionism copied from the developed territories. Thus, for example, encouragement is given to trade unions while legislation is approved requiring the payment of minimum wages, specified conditions of work and the setting up of tariff barriers. Foreigners are often forbidden to work in particular societies despite the skills and the aptitudes which they may be able to bring, and this particular restrictive practice, which is largely a relic of tribal xenophobia, frequently commands a very wide range of popular support. All this means that developing communities find the struggle towards economic development more difficult than their counterparts did in the developing world of the last century and their task is thereby rendered more arduous while progress is correspondingly retarded. The long-term geographical implications of this situation are very considerable indeed.

Domestic economic progress can take place in a hitherto traditional society only as a new entrepreneurial class emerges, but it is very difficult for such a class to develop against the opposition of strongly entrenched social groups. In general, for example, landowners have not been great innovators because of the possibility that the ensuing economic and social convulsions might deprive them of their land, and with it their economic and social prestige and power. This explains why, in South America for instance, revolutions have formed an essential prelude to economic progress. It has been argued that frequently the new entrepreneurial class has begun as powerful personalities or groups, often, though not always, from the lower strata of society and sometimes originating in an underprivileged alien community, found any form of social uplift barred to them and who therefore turned towards economic activities as an alternative. The term "sub-dominant group" has sometimes been applied in such instances. Possible examples of sub-dominant groups include the Nonconformists of eighteenth-century England, the Chinese in south-eastern Asia, the Lebanese and Syrians of West Africa, and the Indians in East and South Africa.

It is important to note, however, that the emergence of such an entrepreneurial class has typically been in trade or in industry; rarely has it been in agriculture with the outstanding exception of the landowners who initiated the Agricultural Revolution in England. A small number of determined men, however, can exercise powerful effects in trade or industry and one out of all proportion to their numbers, but the same is less true in agriculture or pastoralism where the producing units are comparatively numerous and more dispersed. This is perhaps one main reason why agricultural and pastoral societies typically remain wedded to traditional ways.

A related factor which can greatly hinder economic development is the widespread apathy which is so typical of traditional societies, combined with a general lack of interest in achieving what the western world would describe as a higher standard of living. The apathy is not surprising when we consider the poverty, malnutrition, ignorance and the constant battle against natural hazards which have been the rule in primitive societies and the lack of interest in raising living standards in part stems from this. Other factors, however, are involved and we shall pay attention to some of these later.

Other social features also exercise a marked influence upon economic growth but their influence is often a long-term one and not easily or exactly demonstrated. Education is such a factor. Of its importance to a developing society there can be no doubt, for without it there will be none of the skills upon which an advancing technology depends. There is much in favour of the arguments that education, to be maximally effective, needs to be adapted to the technological and administrative, as well as to the cultural and academic needs of a country, and that it should not be the means of creating a disgruntled and largely unemployed intellectual class. Education can have the important effect of lessening the effectiveness of established taboos and restrictive habits, examples of which we have seen above, but on the other hand an over-rapidly expanding educational sector can be a serious financial burden for a developing society to bear.

CLIMATE AND THE HUMAN FACTOR

The main facts regarding the climates of the humid tropics were outlined in Chapter I, while the effects of climate upon activities such as agriculture receive attention elsewhere in this book. At this point we are concerned with the direct effects of climate upon man himself. We have already mentioned the widely prevailing belief that climatic conditions in the humid tropics are unfavourable to man because the constant, or at least the frequent, combination of great heat and humidity produce ill-health, laziness and inefficiency and is often associated with mental and moral degeneration.

It must in fairness be conceded at this point that there is some truth in these beliefs. Constant heat and humidity do sap vitality and there can be few, if any, white people who have actually lived and worked in the humid tropics for considerable periods of time who would dispute this. There is indeed evidence to show that even indigenous inhabitants are unfavourably affected by climatic extremes such as those which occur in the humid tropics.

We are, in fact, simply dealing with a specific instance of the generalisation that man does on the whole find most favourable climates which are equable and temperate, though seasonal variations are advantageous.

But any wide departure from this "norm" brings some measure of discomfort if not actual hardship, particularly when the variation has to do with temperature or precipitation. Man does not generally find areas of climatic extremes hospitable though to varying degrees he has learned to cope with them. Various attempts have been made to quantify the whole problem, one of the earliest being that of Huntington (1915) who believed that for physical work the optimum condition was an outside mean daily temperature of about 18°C (64°F), and for mental work 4·5°C (40°F). An ideal relative humidity is 75–80 per cent. Great seasonal extremes of temperature are not helpful, he believed, though some variations are desirable. Brunt (1943), however, has argued that optimum conditions for a healthy and active life include a mean temperature not exceeding 24°C (75°F) in the hottest month and a general mean of 19°C (67°F), with a relative humidity of about 50 per cent.

Griffith Taylor, who spent much time in Australia, put forward the idea of a "comfort frame" in his work on that continent (first published 1940). For him ideal conditions obtain when wet bulb temperatures lie between 7° and 13°C (45° and 55°F) with an upward extension to 15·6°C (60°F) when conditions are rarely uncomfortable. Above 18°C (65°F) conditions are said to be often uncomfortable, and above 21°C (70°F) usually so. Relative humidities should range between 70 per cent near the upper limits of the comfort temperatures and 80 per cent near the lower limits.

It is clear that the voice of authority on this matter has a somewhat uncertain ring though there is common agreement that in general terms the humid tropics are too hot, too humid, and too monotonous in character to offer anything like an optimum environment for human life and activity. This, at least, will be clear from the figures given above. But the matter is not as straightforward as all that. Elevation and aspect, for instance, play leading parts in the humid tropics in determining levels of comfort. The role of elevation is well enough known and it is also a matter of common knowledge that where other circumstances, for instance political, have been favourable white people have settled on high ground in the tropics, especially in Central and South America and in East and Central Africa. The importance of aspect, however, may be less widely appreciated though inhabitants of the tropics themselves are well aware of it for they well know that air movement usually produces a cooling and refreshing effect which can be very marked. This comes about simply because moving air speeds up the rate of evaporation of perspiration from the body and it is the resulting intake of latent heat from the surface of the skin which results in the cooling effect. This effect is most pronounced when a person is exposed to a breeze blowing through an airy, well-ventilated room but it is noticeable if the air is artificially stirred by an electric fan or even by the old-fashioned punkah. It follows that aspect can play a tremendously important part in house siting; it is by no means unknown in the humid tropics for two houses situated not far apart to experience completely different micro-climates as one is well placed to benefit from the prevailing breezes while the other is not.

Two qualifications must be introduced at this point. The first arises when a long dry season is experienced. When this is the case it becomes important to make provision against the greater temperature range, against the strong dry winds which commonly blow during the dry season, and against the large amounts of dust typically blown along by these winds. The open house structures typical of the equatorial lands with their thin palm "tiles" and walls made of interwoven strands of vegetation or of wood disappear under these conditions. Instead, indigenous houses have thick mud walls, small apertures (windows and doors) and they are heavily thatched. By these means wind and dust are excluded as far as possible and fairly even indoor temperatures maintained. The second qualification has reference to more sophisticated homes which are air-conditioned. In such homes, too, windows should be smaller than in the indigenous open house described earlier, while rooms should normally be fairly small. Through draughts should be excluded.

There does seem to be general agreement that climatic conditions in the humid tropics are such that careful measures are necessary to ameliorate the worst extremes and for a reasonable degree of comfort to be achieved, but this is not something peculiar to these latitudes; corresponding measures are frequently essential in other parts of the world, though naturally they are of a different character. Air-conditioning and the construction of suitable types of houses need not be more of a burden in the tropics than elsewhere as will be abundantly clear if we consider, for example, the vast sums of money now spent in "milder" climates on central heating, double glazing and cavity wall insulation.

As a final point we may note that just as extremes of climate elsewhere can produce troublesome bodily effects, so they can in the tropics; we need think only of the painful skin sores and cracks which can be produced by drying winds such as the harmattan of West Africa or the winter monsoon of India, or of the mercilessly irritating prickly heat which occurs on the skin in the form of a rash as sweat glands break down under excessively warm and humid conditions. There comes a point when evaporation of bodily perspiration cannot keep pace with its formation and it is then that the prickly heat rash appears. At the same time, however, we cannot but admit that the responsibilities for many of the troubles earlier attributed to the climate must now be laid fairly at quite different doors; these include disease, unsuitable or inadequate diets, the wearing of unsuitable clothing, and an inability on the part of migrants to adjust to alien cultures and patterns of living.

DISEASE

In recent years there has been a marked shift of informed opinion regarding the status of disease in the humid tropics. It is a somewhat similar shift to that which we have just examined in the case of climate which was originally thought to constitute an insuperable bar to permanent immigrant settlement and which was held to have the effect of a kind of natural blanket, muffling and stifling human effort. We now see that such a view is untenable, at least in its cruder forms. A similar change has come over our thinking with regard to the status of disease for we now have at hand the means to combat this very real body of scourges. This is not to say that earlier thinking on this matter was wrong; in the context of the times it was entirely correct to see in the heavy incidence of disease in the humid tropics an impassable barrier to many forms of effort and development; it was not, after all, for nothing that parts of West Africa used to be known as the White Man's Grave. A case is recorded of a certain despatch from Sierra Leone, that before it reached London the officer who composed it, the clerk who drafted it, and the Governor who signed it had all died! Today, however, we possess far greater knowledge on the subject of disease and we have means of dealing with the problem which our ancestors could never have imagined. It is frequently forgotten that life in the present developed countries was not particularly healthy many years ago when the average expectation of life was less than 30 years, though, significantly, this could not be attributed entirely to the incidence of disease. It is only as we have come to understand the nature and causes of disease that we have been able to attack it and, in some cases to eradicate it.

In a sense, therefore, it is possible to argue that the dominance of disease is a passing one, but even so we must admit that at present the problem remains very much with us in the humid tropics where the combination of high temperatures and high humidities offer ideal environments for an astonishingly high number of disease vectors. Prominent tropical diseases include yellow fever, malaria, cholera, filariasis, trypanosomiasis, various intestinal diseases such as bilharzia, diarrhoea, amoebic and bacillary dysentery, tetanus, ankylostomiasis, elephantiasis, ... the list could go on. Gourou (1968) states that the intestine of an inhabitant of Yucatan when seen under the microscope has the appearance of a "museum of horrors" in which there are so many unwelcome organisms that immunological prophylaxis would appear to offer no remedy. It is a matter of the slow learning of hygienic habits together with improved living conditions.

In addition to the list of diseases set out above we should note that many, perhaps most of the complaints met with in temperate lands are also known in the tropics. The incidence of tuberculosis, indeed, is said to be increasing in tropical countries *pari passu* with the modern tendency to congregate in towns, frequently under conditions of serious overcrowding, while measles, which is not viewed as a dangerous disease in cooler lands, attacks the inhabitants of tropical lands with a savage virulence which can prove fatal.

Much remains to be done in the field of preventative medicine and the issues involved are complex. In some instances disease can be controlled by vaccinations, injections, or by the systematic taking of a prophylactic, while the careful boiling and filtering of all water used for alimentary purposes, though troublesome, is essential to the maintenance of good health. The constant wearing of some kind of protection on the feet both indoors and outdoors can also at times be a nuisance, but it is imperative, particularly on wet ground, while some form of protective clothing, especially after dark when mosquitoes are most active, is also important. Bathing or swimming in streams and rivers can be highly dangerous. These are, of course, personal measures against many health hazards. On a community scale other steps to improve the general health of the society should be taken. These may include the spraying of extensive areas to eradicate the dreaded mosquito; Ceylon has shown that in this way the incidence of malaria can be very greatly reduced for as a result of the use of DDT against endemic malaria the death rate fell from 22 per thousand in 1945 to 10 per thousand in 1952. Between 1946 and 1947 alone the figure fell by 34 per cent.

It is not proposed at this point to give morbid descriptive accounts of the various diseases. Such accounts can hardly be classed as geographical and they are available elsewhere. That such diseases carry geographical implications, however, there can be no doubt and by way of illustration of this we might well examine the example of malaria as a case study. Much of the following is based on Prothero (1965).

A Case Study: Malaria

Malaria has been a widespread disease in the world from very early times and one which is debilitating as well as being directly and indirectly responsible for a very large number of deaths. Some authorities have claimed that it was one of the factors which led to the decay of the Roman Empire when its incidence increased following the neglect of drainage schemes (on the Pontine Marshes, for example) and the consequent accelerated breeding of the *Anopheles* mosquito upon which the propagation of the disease depends. It was not in fact until the close of last century that it was shown that malaria is due to a parasite which the female mosquito takes in with the blood of an infected person. The parasites subsequently develop within the mosquito before being passed on into the bloodstream of another person from whom, in turn, after ingestion and development, the infection can spread further. Both warmth and fairly copious supplies of water in which the mosquito can breed are needed for malarial infection to persist; the parasites developing within the body of the *Anopheles* die if they are exposed to temperatures below $15 \cdot 5°$ C ($60°$ F) for several nights in succession, while the mosquito itself needs temperatures above that level for itself and its larvae to survive.

It is in some ways unfortunate that different species of *Anopheles* flourish under slightly differing physical environments, as this means that action taken against one species may leave another untouched. Thus, for example, some prefer clear, shady water as a habitat while others prefer clear, running water warmed by the sun. The Red River delta in south-east Asia is usually free from malaria because the *Anopheles* cannot tolerate the muddy waters of the rice fields, but the adjacent hill country with its clear, running streams is highly malarial. These streams are today warmed by the sun following deforestation of the slopes and the destruction of the trees which formerly shaded the water.

The distribution of malaria throughout the inter-tropical lands generally is widespread except in areas such as the Andes which are too high and therefore too cold, and in other areas such as most of the Saharan, Arabian and Australian deserts which are too dry, while few parts if any of the humid tropics are free from it. The disease varies considerably, however, in its virulence, very largely in accordance with the type of mosquito vector which is dominant. For instance, in Brazil the disease is widespread, as we should expect, but the resident vector,

Anopheles darlingi, is very exacting as regard its physical environment and it is not one of the more dangerous varieties. In the 1930s, however, the *A. gambiae*, a far more virulent vector which is endemic over most of tropical Africa, was accidentally introduced into the areas around Natal, and almost immediately a catastrophic epidemic of malaria broke out. By 1938 nine-tenths of the inhabitants of the Natal region were infected and in one region 20,000 out of 100,000 patients died. Fortunately the Natal region experiences a severe dry season annually and it proved possible to exterminate *A. gambiae* by attacking it during this part of the year.

One major difficulty facing any malarial eradication campaign is that many inhabitants of the tropics have come to accept the disease as part of the natural order of things; indeed, in many illiterate communities it may hardly be recognised as a disease at all! It is not easy to convince such people that preventative and eradication measures are necessary, and this attitude can seriously jeopardise the chances of a successful campaign.

The road to effective control and final eradication of malaria is not an easy one. It is true that drugs are now available which will cure those suffering from the disease and which will prevent its being contracted at all, but these are expensive, while the curative drugs will not prevent re-infection. There is no drug which will give long-term immunity, neither is any form of vaccination or injection at present available for this purpose. In addition medical facilities in most countries of the humid tropics are very limited and large numbers of inhabitants are not reached by them.

At present, therefore, the only possible avenue of attack is that directed against the mosquito vector responsible for the disease. If the number of vectors can be greatly reduced so that the transmission rate is brought down to a very low level and is held down for several years then the disease will naturally die out. Even if the measures are then reduced and the mosquito population increases again the disease will not recur *unless*—and this is a vitally important *unless*—the disease is re-introduced. This, of course, can easily happen through population migration from infected areas such as might occur if workers move between their homes and developing areas where paid work is available. This scheme of things is illustrated in Fig. 10. It is significant that where an eradication programme has failed the failure has generally been due to the human factor and particularly to population mobility. It must be admitted, however, that the spraying coverage is sometimes not complete, usually because of incomplete knowledge of the region concerned.

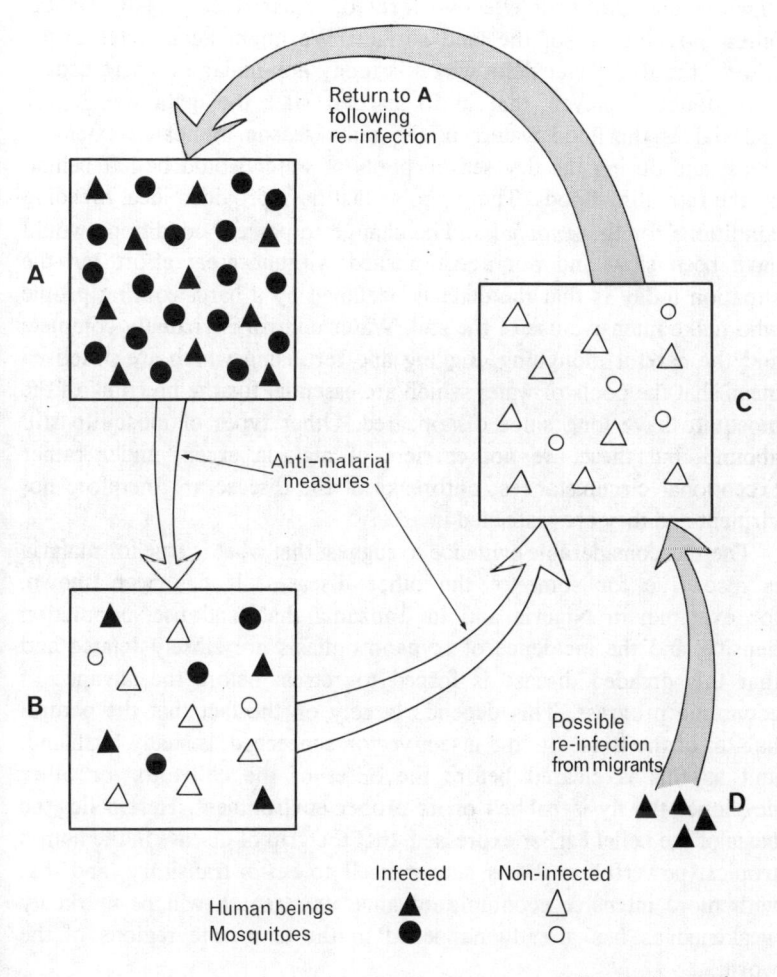

[Based partly on Prothero

FIG. 10.—Malaria control.
    A. A malarial area, inhabitants and vectors infected.
    B. The same area following eradication measures. Fewer infected vectors, the number of non-infected inhabitants increasing.
    C. The same area, infection has disappeared.
    D. Possible movement into the area of infected persons. If not checked, the new infection could spread, with a consequent return to stage A.

There is a considerable body of information now available which suggests that the most effective form of malarial control lies in the fullest possible use of the land surface. We might again refer to the case of the Red River delta which is today non-malarial. There can be little doubt, however, that in its natural state the delta was highly malarial as the flood waters of the rainy season inundated extensive areas, and during the dry season pools of water would be left behind by the retreating floods. These pools must have provided ideal breeding conditions for the *Anopheles*. The change to present conditions would have been slow and not accomplished without great effort but the situation today is that the delta is occupied by a hard-working people who make intensive use of the soil. Water control is virtually complete and the careful ploughing, digging and fertilising which are practised mean that the pools of water which are essential for the breeding of the mosquito have long since disappeared. Other types of mosquito still abound, but these are not carriers of malaria except under rather exceptional circumstances; outbreaks of the disease are therefore not virulent and they are restricted in area.

There is considerable evidence to suggest that what is true for malaria is also true for some of the other diseases. It has been shown, for example, in Nigeria and in Tanzania that land use, population density, and the incidence of trypanosomiasis are closely related and that this dreaded disease is forced to retreat before the advance of economic progress. This depends largely on the fact that the natural habitat of the tsetse fly, the insect vector concerned, is shady bushland, and as this is cleared before the onset of the cultivator or other developer the fly is robbed of its proper environment. Herein lies the basis of the belief earlier expressed, that the grip of disease in the humid tropics, powerful as it has shown itself to be, is transitory, and that with more intensive economic advance its strength will be markedly weakened as has already happened in the temperate regions of the world.

DIET AND NUTRITION

It is generally believed that malnutrition is both chronic and widespread in the humid tropics, a belief which was referred to in Chapter I. While emotive statements on this subject by their very nature receive considerable attention their validity is often open to considerable doubt and it is difficult indeed to arrive at a balanced assessment of the situation. One reason for this is the simple fact that the scientific measurement of man's food requirements is something which has begun comparatively recently and the task is far from complete; this makes such dogmatic statements such as that issued by Lord Boyd Orr (p. 2 above) doubly unfortunate for they can give an illusion of certainty which is entirely spurious.

Food, if we leave aside the matter of pleasure in eating (which in fact is a very important consideration but one with which we can hardly deal here), fulfils a number of purposes upon which the bodily, and indirectly the mental and spiritual, health of a person very largely depend. One such purpose is the provision of the energy which maintains bodily warmth and activity. Most foods contribute in some measure to this essential function, the unit measurement of the energy received from various sources being the calorie, and it is upon our daily intake of calories that we depend for our bodily activity and energy. But the body must also be maintained in a good state of repair and the most important compounds serving this purpose are proteins, which are nitrogenous compounds found in complex organisms. Animal matter, weight for weight, tends to contain more readily assimilable protein than vegetable matter, and it follows that a person who depends mainly or wholly upon vegetable proteins needs to consume a greater bulk of these foods.

Other food requirements include the vitamins which are essential for the prevention of certain diseases and for the maintenance of good health, and also various mineral "trace elements" analogous to those needed by growing plants; the actual amounts of trace elements needed, though vital, are very small. It is important to note (Clark and Haswell, 1970, 4) that shortages of vitamins and minerals are not very likely to arise in primitive communities because the food consumed, limited in amount as it may be, is normally derived directly from natural sources. A vital deficiency can be induced, however, by any interference with this natural food supply, as in the now notorious example of the polishing of rice before sale in certain wage-earning communities of south-eastern Asia. It was the outer layer of the whole grain which contains the essential vitamin B_1 which was removed by the polishing, and as whole rice was replaced on the market by the polished variety there was a correlated increase in beri-beri among the poorer members of urban and industrial groups.

Among the efforts made to quantify essential food inputs are those which attempt to specify the minimum daily calorific intake per person. A figure often quoted is 10,460 joules, while figures calculated by the FAO are 13,400 for a man and 9630 for a woman; other writers believe that a minimum of 12,560 calories *per caput per diem* irrespective of sex is essential for any marked output of energy and initiative.

On the other hand since the metabolic rate is lower in tropical than in temperate lands it is suggested that lower intakes should suffice in low latitude regions, and in Malaya (Hodder, 1973, 72) the recommended minimum intake for a peasant performing normal outside work has been put as low as 8790, which seems a very low figure. Requirements of calories vary considerably, however, in different circumstances. For instance they are related, though not proportionally, to the weights of the individual bodies concerned and also to environmental factors, as we noted above. Calorie requirements are in addition increased by active physical exertion (the number of hours worked per day is obviously important in this respect), and in women by pregnancy and lactation.

It is therefore necessary to view the various attempts which have been made to specify essential calorie intakes with considerable reserve, and when further attempts are made to measure specified amounts against actual intakes we must remember that it is unlikely that figures obtained are accurate. In any case it must be borne in mind that an average figure being what it is the calorie intake of large numbers of people must be below any statistical mean put forward in respect of any large community. Another reason for noticeable variations is environmental; Clark and Haswell (1970, 20) quote the results of a survey undertaken in India in twelve villages distributed over areas of four different soil types. The following table shows the average recorded *per capita* consumption of calories per day in the differently situated settlements.

In villages situated on:	Average energy intake (joules)
Well-drained loam with irrigation	8961
Stiff black clay	8446
Poorly-drained clay	8392
Poor sandy soil	7756

It is clear that an overall average figure which conceals such variations is of somewhat limited value.

One point which is beyond dispute is that the diets of most of the inhabitants of the humid tropics are monotonous to a degree rarely if ever experienced in more developed lands. There is an overwhelming dependence upon cereals, for the most part rice (especially in south-east Asia), millets, maize or wheat, and also upon certain root crops, notably potatoes, yams and cassava. Consumption of sugar is low but is probably increasing while that of meat, dairy produce and fats is also low except in one or two special areas. The importance of this set of facts is considerable bearing in mind what was said above (p. 71) regarding a vegetarian diet.

Pedler (1955) has given some very interesting details regarding the relations between dietary deficiencies and work capability among West

Africans working on the scheme to irrigate parts of the inland delta of the Niger above Timbuktu (Jarrett, 1974a, 274) where, it seems, new workers did not have the physical strength to carry out the various tasks allotted to them. Their strength could be built up, however, by giving them meals containing meat—which some of them had never eaten before! During the first week they were given meat on 2 days and this allowance was increased by 1 day in each subsequent week until by the sixth week they were eating meat each day. By this means it was found possible to increase their work load from one-third of the normal during the first week to a full load from the fourth week onwards. A more recent example quoted by Sampedro (1967, 32) concerns the construction of a highway in Costa Rica which was making very poor progress until new contractors increased the labourers' rations. Until that happened the workers, of whom 70 per cent were Costa Ricans and 30 per cent were from the U.S.A., were managing to move only 240 cubic metres of earth per man per day. With the improved diet and although the proportion of Costa Ricans increased to 88 per cent the figure went up to 1150 cubic metres.

A great deal of thought has gone in recent years into the question of what can be done to "solve" the "problem" of world hunger, though whether such a situation is in fact a problem capable of "solution" may be open to doubt. In the long run it seems reasonable to expect that increasing production will yield sufficient additional food supplies to provide for the populations of the humid tropics as has already happened in the developed countries, but meanwhile the situation is serious though almost certainly it is not as bad as many authorities have argued. Clark and Haswell (1970) believe that the overall deficit today is probably of the order of 15–20 per cent and they completely reject the figure given by Boyd Orr quoted earlier (p. 2) which was based upon data of very questionable validity.

Even so there is no room for complacency and various suggestions have been put forward to deal with the short-term situation; these include redistribution of food surpluses and the better use of existing resources. The first possibility has frequently been urged and there is no doubt that very considerable food surpluses have existed in certain countries and in certain commodities in the past. It is perhaps unlikely, however, that this situation will be repeated, at any rate on any large scale; the present world situation appears to be one not favourable to the accumulation of stocks of surplus food in the face of increasing population and affluence except, possibly, under artificially contrived conditions such as exist within the E.E.C. with its butter and and beef "mountains."

The second possibility mentioned above, the better use of existing resources, offers more potential. There is some force in the argument that in many areas more nutritive crops could be grown than is at present the case and that better use could be made of crops at present grown. For instance on better quality land it may be desirable to grow crops like potatoes or rice where this is possible rather than fodder crops for these crops will provide far more joules per unit area *per annum* than will livestock. On the other hand stock can often graze very successfully on poorer pastures where crop growing would not be very successful because of poor soils. Cassava, which is widely grown in the humid tropics, has only a 1–2 per cent protein content but many pulses carry between 25 and 45 per cent, while these leguminous crops frequently can enrich the soil in which they are grown. It is also worth noting that cassava leaves have a 12 per cent protein content and it is technically possible to extract this protein and to produce from it a perfectly palatable food.

The fuller use of natural fauna is also a topic worthy of attention, for experiments in Africa suggest that there are real possibilities of exploiting natural game for their meat. On the poorer savannas the herds of Thompson gazelles, zebra, wildebeest, eland and impala are large and there is the important point that animals such as these browse upon different plants from those eaten by cattle. Economically it may make good sense in many instances to keep poorer grazing land as pasture for wild life from which meat can be obtained. There are also great possibilities for protein production in inland waters. Fish farming, which is also being developed in some parts of West Africa, provides an important source of protein in south-east Asia and experience has shown that very high fish yields can be secured with the help of appropriate fertilisation measures in warm tropical inland waters.

POPULATION

One of the most startling features of our time is the unparalleled increase in population which is taking place in the world as a whole. It has been calculated, admittedly from indirect sources, that in the first century A.D. the total world population was probably about 400 million while by 1650 it had modestly risen to roughly 500 million; in 1750 it may have reached about 750 million. Since that time, however, it has risen in an astonishing manner while, more alarmingly, the rate of increase itself has gone up from about 0·3 per cent between 1850 and 1900 to 0·9 per cent between 1900 and 1950 and to 2·1 per cent between 1950 and 1965. The estimated rate for the latter part of this century is between 1·8 and 2·5 per cent. Actual world population had increased to 1000 million by 1850, to 2000 million in 1930, to 3000 million in 1960, and possibly to almost 4000 million by 1975. And if we assume an annual growth rate of about 2 per cent it is likely that there may be no fewer than 7000 million people living on the surface of this planet by the year A.D. 2000. An annual growth rate of 2 per cent does not sound very formidable but in fact it will lead to a doubling of any initial amount after only 35 years.

The World Bank (Finance and Development, Vol. 10, No. 4) has published population projections as far as the year 2000. The projections are three-fold: A, assuming that fertility rates remain at their present level; B, assuming a moderate fertility decline; and C, assuming a "fast" decline (a falling of fertility rates by between 1 and 3 per cent annually until replacement level is reached. This could halve the rate in 30–40 years). The projections are as follows.

Year		N. America	Latin America	Africa	Europe & U.S.S.R.	Asia & Australasia	World
1970		226	280	346	704	2051	3607
1980	A	250	378	457	760	2596	4441
	B	249	372	534	757	2558	4389
	C	248	367	451	754	2521	4341
1990	A	280	517	619	815	3377	5608
	B	275	491	597	809	3196	5368
	C	271	465	578	801	3039	5154
2000	A	308	710	861	870	4447	7196
	B	297	635	783	854	3968	6536
	C	288	567	677	843	3583	5958

The humid tropics are, of course, sharing in this population increase, though with very uneven results because present population densities vary so greatly. This is illustrated by the table on p. 10 above. In fact it is tropical territories which are very largely responsible for the present world demographic situation, as the table above suggests and as Fig. 11 shows, because it is particularly these countries which record above-average rates of growth, while on the whole the territories of higher latitudes show lower-than-average rates.

The main reason for this is not difficult to find; it lies in the enormous expansion in medicine and in public health measures which has occurred during the present century which is itself founded upon an ever-widening available resource base, particularly of food supplies. Thus, while expectation of life in the L.D.C.s in the early 1940s was 30–35 years, today, outside tropical Africa it is much higher:

Rates per cent

Above average
>3·0
2·0–2·9 World average 2·0

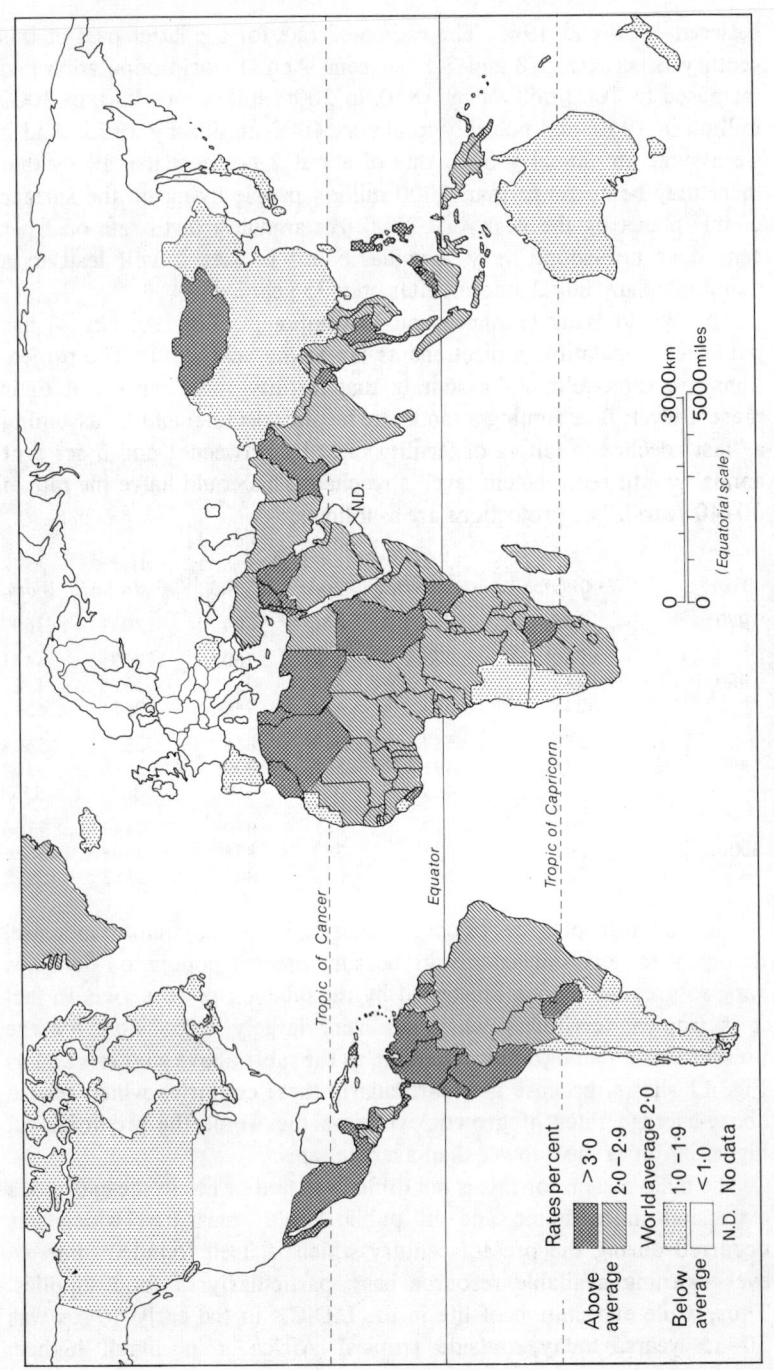

Below average
1·0–1·9
<1·0
N.D. No data

0 — 3000 km
0 — 5000 miles
(Equatorial scale)

FIG. 11.—The world: natural population increase.

average figures which are known with some certainty, for instance, include Mauritius (59 years (males), 62 years (females)), Jamaica (63, 67), Sri Lanka (62, 61), Western Malaysia (63, 66) and (for comparison) the United Kingdom (69, 75). This trend towards higher life expectancy began in Britain during the second half of the eighteenth century, and from there it spread to the rest of the industrialised world; today most countries are affected by it. It was based partly upon higher standards of living thanks to the tremendous advances which were made in industry and agriculture, but more directly it was due to the dramatic fall in the death rate. This is illustrated in Fig. 12 which

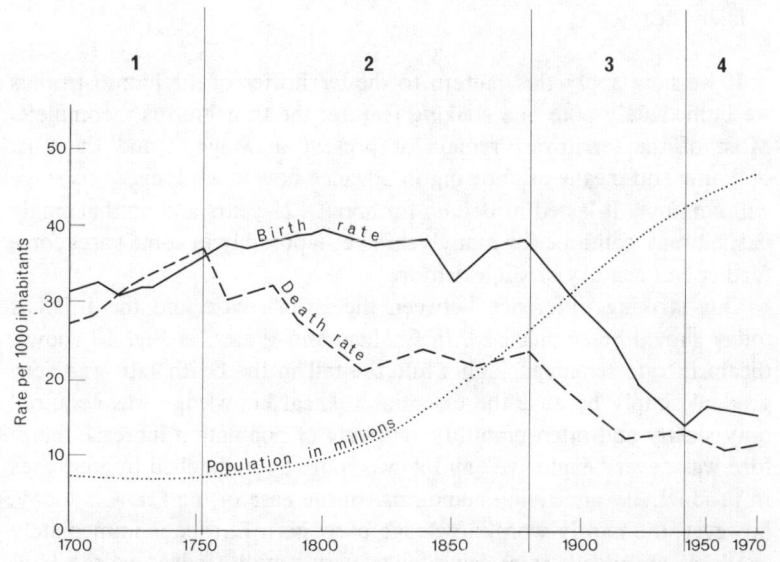

FIG. 12.—Demographic transition in England and Wales (1700–1970). For comments *see* text.

recognises four stages in this period of rapid demographic transition. Although the diagram has special reference to England and Wales it can stand in its own right as a model to illustrate the principles involved and it may with modifications have reference to any territory. The four stages are as follows.

1. *Early fluctuating stage.* Both birth rate and death rate are high but on the whole the former is slightly above the latter. There is therefore a very slow population increase which is at times interrupted, for instance by warfare, plagues and other epidemics which result in temporary but noticeable population decreases.

2. *Early expanding stage.* The death rate falls rapidly for the

reasons already stated but the birth rate remains high. Population therefore begins to increase rapidly.

3. *Late expanding stage.* The death rate continues to fall but at this stage the birth rate also decreases. Population therefore continues to increase but at a slackening rate.

4. *Late high steady stage.* Both the birth rate and the death rate are now comparatively low but as the former is slightly the higher population continues to increase though at a noticeably reduced rate. Plagues and epidemics generally are not the powerful killing agents which they were in Stage 1 and sharp fluctuations such as were typical of that period are not very likely to recur. Growth is therefore fairly steady.

If we now apply this pattern to the territories of the humid tropics we immediately note one striking feature: the transition is incomplete. Most of the territories remain at present in Stage 2 and there is, of course, no means of knowing in advance how much longer this stage will continue. It lasted in Britain for about 125 years and on that single basis it may continue for many years yet—possibly in some cases for a further half century or slightly more.

One striking difference between the British case and the L.D.C.s today should be emphasised. In England and Wales, as Fig. 12 shows, the birth rate remained high while the fall in the death rate was very gradual simply because the essential medical knowledge was acquired only slowly and often painfully. The rate of population increase therefore was never "explosive" and it was more than matched by increases in productivity and living standards. In the case of the L.D.C.s today, however, the hardly-won knowledge of western Europe is immediately available and death rates have fallen very rapidly indeed as the Malthusian checks of an earlier age have been dramatically reduced in force. For instance the incidence of smallpox, leprosy and malaria, to name only three scourages of formerly world-wide distribution, is much reduced today; we are, in fact, approaching the stage where smallpox has been totally eradicated from all but a few pockets in the world as a whole. The extended provision of piped drinking water in many areas, including much of India, has done much to eradicate water-borne diseases. The older checks of famine and war have also been much reduced in modern times as it becomes possible to distribute food supplies more efficiently, especially to areas of acute shortage, while the *Pax Britannica* put an end to much local warfare—though this may prove to have been just a temporary reprieve. These are the root causes of the population explosion which we are witnessing today for they mean that in the L.D.C.s birth rates remain high as in Stage 2 while death rates are more comparable to those of Stage 4 of our diagram.

A population increase such as that now taking place in the world has repercussions of the greatest importance for the world as a whole as well as for the individual territories concerned, and we should now turn our attention to some of the relevant points.

In the first place the increase has enormous importance with regard to food supplies, not only within the L.D.C.s themselves but also in the world as a whole. It is obvious that very large increases in food output will be needed simply to feed the expanding populations of the various countries at present levels—which in many cases is very low. The FAO (1963, 72) have estimated for example that Africa will need to produce more than $3\frac{1}{3}$ times as much food in the year 2000 as it did at the end of the 1950s; corresponding figures for Latin America and the Far East are about $3\frac{1}{2}$ and $4\frac{1}{4}$ respectively. These are very formidable figures and they clearly represent a task of very considerable magnitude though we should be under no illusion what they mean when the World Bank can report in 1975 that about 650 million inhabitants of the world live on miserable incomes of less than £24 *per annum*. With regard to the world as a whole the problem of food shortages is equally pressing and it has become a very serious issue for the industrial nations which rely heavily on imported foods. Signs that it is making its impact felt are not difficult to find, notably, of course, in the generally increasing world prices of foodstuffs.

The second point to notice is that the problem presented by rapid rates of population increase is far more serious today than ever it has been in the past. This is largely because when the present escalation of world population began in what are now the developed territories the populations of these territories were comparatively small (*see* for instance Fig. 12). Where initial populations are low even fairly high rates of increase will not quickly result in large actual increases. We must recognise that this is even today the situation in many parts of the humid tropics, but very considerable numbers of people live in some of the territories concerned (*see* the table on p. 10 above), and high rates of increases in these instances result in very high actual population increases. This is the situation with which we are faced today and in many respects we must allow that it is a frightening one.

Annual rates of growth in selected territories over the period 1963–69 include 2·2 in Mauritius, 2·4 in Jamaica and Ceylon, 2·7 in Thailand, 2·8 in Western Malaysia and 3·5 in the Philippines. These figures contrast starkly with that of 0·6 per cent for the United Kingdom. The 1971 census of India shows that the population of that vast territory rose during the preceding decade from 439 million to 548 million, an increase of 24·6 per cent compared with 21·5 per cent during the previous decade and only 5·7 per cent during the first decade of this century. If the present rate of increase is maintained it is likely that the population of India alone will reach 1000 million by the end of the century.

Population structure in the territories of the humid tropics is of course profoundly affected by such increases, a point illustrated in Fig. 13. A very large proportion of the inhabitants of these areas is today below 15 years of age. The figure for Africa as a whole is 42 per cent, with only 4 per cent of the total population over 65 years, while for Latin America it is 41 per cent (the figure for Venezuela is as high as 55 per cent) with 3 per cent over 65, and for south-eastern Asia

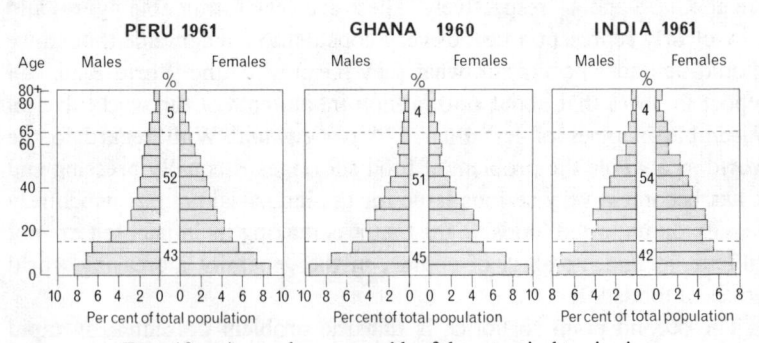

FIG. 13.—Age and sex pyramids of three tropical territories.

it is 40 per cent (3 per cent over 65). These figures contrast with 25–30 per cent for the more-developed territories with up to 10 per cent over 65. The situation in three selected territories of the humid tropics is shown in Fig. 13 and this important point regarding the very high proportion of young people is clearly shown. In each of the three cases shown well over 40 per cent of the total population is below the age of 15 years and these young people will during the next decade or so reach child-bearing age and have families of their own.

The implications of this situation are very great. The increasing numbers of young people, for example, mean that the communities concerned must bear heavy financial burdens in the provision of social services, including medical and educational facilities for the young, and while such expenditure is greatly to be desired on humanitarian and even moral grounds it is not directly productive. Such a situation is bound to put a developing country under severe financial strain.

It is true that many observers have argued that population increases have stimulated economic and technological progress as they have encouraged farmers and industrialists to seek more efficient methods of production under the stimulus of increased demand and consequent high prices. But as we have suggested earlier there is no real comparison between the comparatively steady population increases which took place in the developing territories of last century and the "explosion" which is taking place in the L.D.C.s today. We can list, by way of summary, several disadvantages of a too-rapid, too-great demographic increase such as we are now discussing.

1. A rapid population growth cannot be matched by a corresponding increase in the available food supply.

2. It also leads to diversion of capital from productive investment to social amenities such as hospitals and schools. These two points have been noted above.

3. A point closely related to 2 above is that population growth inevitably starts "from the bottom," *i.e.* in the lowest age groups (Fig. 13). As attempts are made to feed and support large families money necessarily is drawn into consumption spending with consequent decreases in savings and investment. Thus the supporting of large families in itself must become an economic burden upon the whole community.

4. As the children grow to adulthood they seek work but it is unlikely that work on a sufficient scale can be found and severe unemployment can result. We can take an example of this situation from Kenya which has the high demographic growth rate of 3·3 per cent. If this rate continues the 1973 population of about 11 million will become roughly 30 million by the end of the century. Of the present working population of 3,800,000 only about 17 per cent (just under 650,000) are employed in urban areas or on large-scale farms as wage earners and it is estimated that if the present unemployment problem is to be solved three times as many jobs must be found for wage earners as at present exist, while if the present rates of natural population increase are maintained *an additional* 230,000 new jobs must be found *each year*. One has only to state the problem to see the near-impossibility of successfully coping with it.

5. A further point arises from the above. A rapid increase in the labour supply must mean that less capital will be available per worker unless supplies of capital are correspondingly increased—a most unlikely event. It is this fact which lies behind attempts made in

some territories, notably in India, to encourage labour-intensive activities such as the construction of small-scale dams, even if in theory expensive capital equipment could be used more efficiently. If such equipment were used, however, the amount of labour required would be small and the enterprises concerned would therefore make only a very limited impact upon the unemployment problem. Unfortunately it must be admitted that productivity under these circumstances is likely to suffer and to the extent that this is true economic growth may well be inhibited. But we must remember that productivity *in itself* without correlated economic developments (such as the emergence of a market to absorb the additional goods produced) is likely to prove a burden rather than a help to additional progress.

The Optimum Population

The concept of the optimum population is one which has gained currency among demographers and economists and it is one of which we should take note. In theory the optimum population may be regarded as that number of people which will produce, when working with all the other available resources, the highest *per capita* economic returns. If any given population moves upwards or downwards from the optimum the *per capita* income will fall—in other words, standards of living will fall though not, of course, evenly throughout the community.

The term is not favoured by all authorities because it lacks precision. It is possible to point to territories in the humid tropics, for instance, and to say that their respective populations are above or below the optimum. This simply amounts to saying that they are overpopulated like India, Mauritius or Barbados, or underpopulated like Brazil, Zaïre or Laos, but when we attempt to say more exactly which countries are near the optimum we find that we cannot do so. One important reason for this is that the optimum population is not a fixed quantity; it changes with improvements in technology and method, so in general it is constantly moving upwards—though it might conceivably move downwards if for any reason a community became a less efficient producing unit. It is also true that the most effective population cannot be measured simply in terms of overall size. Questions of age, sex constitution and skills are very important and cannot be ignored. And there is another very important point; while a population may be too great for one purpose, perhaps for agricultural development, it may *at the same time* be too small for another, perhaps for the establishment of industry which requires a skilled work force and an assured market for its products. A market is not by any means the same thing as a population of any given size; the people must have purchasing power before a market can come into existence. This problem can clearly be acute in small territories like Haiti or Sri Lanka. There is much to be said in favour of the argument that static concepts such as the optimum population, underpopulation and overpopulation, are so generalised and imprecise that they can be usefully discarded.

Population and Economic Production

There is little doubt that the central problem presented by the present demographic situation is that of increasing the rate of economic production so that it lies statistically above the rate of population growth. Only in this way can standards of living be raised. This may be most easily accomplished in the case of small and sparsely populated territories where the existing population is too small to permit labour specialisation and the development of large-scale production and where, as many economists would argue, the population is below the optimum. With increasing numbers of inhabitants, a developing economy and an expanding market the standard of living in such a territory can be expected fairly quickly to show a definite improvement.

The case is quite different, however, when we are dealing with a large initial population which is increasing rapidly. Under these circumstances a territory may well fall a victim to the "low-level equilibrium population trap" discussed by several writers which is shown in diagrammatic form, slightly modified, in Fig. 14. The vertical axis records percentage increases of population and incomes, the horizontal represents the passage of time.

The pattern shown on the diagram will probably now be clear. The sequence begins at P, at which point we may assume that modern ways of life are for the first time affecting the inhabitants of the territory concerned. The results of this as shown on the graph are that the population begins to rise after an initial decrease (the curve starts from a negative rate of increase); this may be due to the production of more food and the adoption of modern hygiene and medicine. At the same time incomes begin to increase, perhaps as agriculture commences to move out of its traditional stagnation. As far as A the rate of income increase lies above that of population increase so the standard of living generally is rising (there will, of course, be individual exceptions to this within the community). At A, however, the "trap" begins to operate as beyond that point the rate of income increase is surpassed by that of population increase. In simple language, although production is still

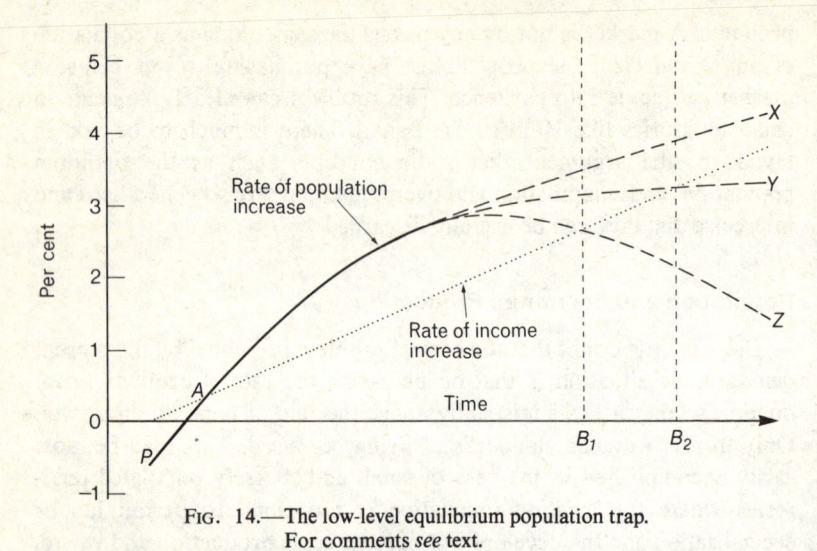

FIG. 14.—The low-level equilibrium population trap.
For comments *see* text.

increasing it is not doing so sufficiently to keep abreast of the increase in population. Even if *per capita* incomes are increasing, as they will be in monetary terms, price levels will be rising even more so that the people generally are worse off than before. This situation will continue until the rate of population increase slows down and is overtaken by that of income increase. If, of course, this does not happen the country may be unable to extricate itself from the trap (situation X in the diagram) but such a situation would at some point become intolerable and it could not be sustained indefinitely.

The actual time when the territory is able to escape from the trap will depend upon the speed at which the rate of population increase falls in relation to the rate of income increase. If there is no actual fall but simply a slackening in the rate of population increase (situation Y in Fig. 14) the escape from the trap will take place at B_2 but if there is a fall (situation Z) it will be brought forward to B_1.

POPULATION CONTROL

It remains for us to ask what are the most effective demographic measures which can be taken to increase the well-being of the inhabitants of the humid tropics. Many possibilities have been suggested, one being that of the redistribution of existing populations. This can be effected in various ways, one being migration to sparsely populated areas thus relieving the pressure on densely populated regions. In

Indonesia, for example, there is official encouragement for inhabitants of the densely populated island of Java to migrate to one of the more sparsely peopled islands, and many Javanese have in fact moved to Sumatra and Borneo. The numbers of migrants involved in such an exercise, however, must remain comparatively small while it has the serious disadvantage that such migration frequently distorts the natural age and sex structure of the donor community. Hodder (1973, 92–3) claims that the migration of Cabrais into central Togo achieved this result and left behind in the Cabrais homeland a striking dominance of females in the 15–59 age groups (120 females to 100 males). Such an imbalance can have very undesirable social as well as economic effects because of the loss of so many members of the virile and active male group.

It is, moreover, very doubtful whether in fact migration of this kind does do very much to help the situation. Certainly, to take one example, the migration overseas of large numbers of Tamils from south India in the nineteenth and early twentieth centuries to work as indentured labourers on tea, sugar and rubber plantations in the British tropics and in eastern Africa has had very little long-term effect on the population of their homeland.

Some measure of population redistribution will, of course, inevitably accompany normal economic development especially as industry develops and as urbanisation gathers momentum, while agricultural progress can also attract migrants as we saw on p. 38 above and as we shall see later in the case of Brazil. These themes will be examined in later chapters as they are dominantly economic in character.

There is little doubt, however, that the most effective demographic measure which can be taken is to lower the birth rate, and this is a possibility which we should now examine more carefully. In one sense it may be said that there is a simple answer to the problem—that of conception control, more usually referred to as birth control. The likelihood of success of this approach has been considerably increased within the last decade or so by the development of the oral contraceptive which is now widely used in many countries, and, in some cases, by the official encouragement now given to voluntary sterilization as in some parts of India.

While all this is true it must also be said that much doubt still surrounds the whole question. In the first place there is still the question whether such an approach can be really effective in such lands as those of the humid tropics where educational standards remain low and where traditional practices (which will be called "prejudices" by those who disapprove of them) continue to retain a strong grip. It is true that educational standards can be raised and long-accepted attitudes changed, but these things take time to achieve. It also remains true that for many folk conception control raises moral and even religious issues and this therefore poses serious problems in many countries. Even so, despite these doubts and difficulties, we can be sure that the importance of this form of population control will increase.

There is a sense, however, in which the pursuit of this approach to the whole problem is sterile (!) and profitless. It may be far more useful and realistic simply to accept the fact that conception controls are now available as helpful, even vital tools, but to recognise that the real answer to the problem lies in the human mind. After all, as Fig. 12 shows, the fall in the birth rate in England and Wales began long before contraceptives were widely available and this fact suggests that while these aids to conception control are helpful they do not lie at the root of the problem. The important point surely is this, that parents will begin voluntarily to restrict the sizes of their families *as soon as they see some positive advantage in the restriction*. Quite frequently during the early days of an economic revolution the birth rate rises as children are looked upon as essential wage-earners among the poorest groups of workers and their wages are sadly needed (this feature is shown on Fig. 12). But as new opportunities for economic and social betterment arise in the developing community a change sets in. Parents begin to see opportunities of advancement for their children beyond anything which they themselves knew when they were young and they also come to realise that such opportunities are more easily grasped by children of small families. Perhaps it is no accident that the beginnings of the downward swing in the birth rate in England and Wales came closely after the passing of the famous Education Act of 1870 introduced by W. E. Foster during Gladstone's first Ministry. While the provisions of the Act were modest indeed judging by present-day standards it did make possible a very considerable extension of education among the poorer members of the community and in so doing it opened a little wider the door to economic and social betterment.

Here, then, may lie the real key to world population control. As soon as people, especially the poorer, begin to see real possibilities for their children and as soon as they see that they all can begin to share in the fruits of economic progress the problem may well be on the way towards solving itself. In this chapter we have only begun to touch on some of the questions which any consideration of the human resources of the humid tropics must raise but lack of space means that at this point we must draw the discussion to a close. The human situation, however, is not lost sight of in the chapters which follow.

Chapter IV

Agriculture in the Humid Tropics: Shifting Cultivation

THERE is no doubt that agriculture is still by far the most important single occupation of the inhabitants of the humid tropics. Some writers comment upon the general similarity between the farming methods employed over the region as a whole (*e.g.* Gourou, 1968, 31) while others stress the striking regional differences which are met with (*e.g.* Hodder, 1973, 97). There is not necessarily any real contradiction between these points of view for even if a basically similar system did exist over such an extensive region it would be remarkable indeed if variants did not occur in different areas; such variants are likely to be based upon differences in local physical environments as well as upon community practices and customs.

Many writers have commented upon the very high proportion of the working population of the humid tropics which is essentially engaged on the land, either as cultivators or as pastoralists. Even today in some areas more than 90 per cent of the total population derive a livelihood directly from the soil while it is frequently above 65 per cent, a figure which contrasts sharply with those, for instance, of Britain and New Zealand which stand at about 5 per cent in each case. Unfortunately this concentration upon agriculture is not matched by substantial outputs of foodstuffs and basic necessities; indeed, the reverse is frequently the case, and it is to this general situation that we address ourselves in this chapter.

SHIFTING CULTIVATION

It is well known that there is one system of cultivation which is remarkably widespread in the humid tropics, and this is, of course, that generally known as shifting cultivation, though it is known by different names in different regions. Thus we have *ladang* in Indonesia, *bewar* in Central India, *milpa* in Mexico and *roça* in Brazil. Quite frequently it is referred to as *swidden cultivation*, especially in the Philippines and in Thailand though *swidden* is not a term which actually originated in the humid tropics. It is an old English dialect word meaning a burnt clearing.

The main features of shifting cultivation are well known. The cultivator first decides upon a piece of forest which he considers is suitable for crop growing. Often the oldest available "bush" is selected, partly because the undergrowth there will be less impenetrable and partly because the soil will carry a higher humus content and will therefore be more fertile. Sometimes the presence of certain kinds of tree, known in Brazil as *padraes*, or tree-guides, indicates land of superior fertility, while Gourou (1968, 31) alleges that many cultivators in Benin (formerly Dahomey) even taste a pinch of soil to test fertility!

The bush is then cleared with the help of an axe or cutlass (machete) though the biggest trees are normally left as it is virtually impossible to fell them with primitive tools, and in any case they act as windbreaks and help to protect the soil against erosion. Useful trees such as the oil palm or those yielding edible fruits are also spared, and they, too, help to give shade and to guard against erosion. Trees which are felled are chopped through where their trunks narrow, often about 6 ft above the ground, and the stumps are left standing so that the cleared patch as a whole typically wears a remarkably shaggy appearance far removed that that characteristic of the trim fields normally found in cultivated lands elsewhere.

The next stage in the operation is to burn the brushwood which is frequently gathered into enormous heaps for the purpose, and at the end of the dry season the whole countryside may be dotted with enormous fires the smoke from which can throw a gloomy, dirty pall over wide areas. The week or two when the burning is taking place can be an anxious time for the farmers for if the rains are early the clearings may be soaked, burning may become impossible, sowing and planting of crops may therefore be impeded, and the subsequent harvest may be reduced to a dangerously low level. The work of clearing should be so timed that the brushwood has time to dry before the burn but it must not be so early that fast-growing plants have time to regenerate themselves in the clearings before the sowing and planting. It is extremely difficult to gauge these activities correctly in the face of climatic uncertainty (p. 28 above).

There has been much debate regarding the value of the annual burning. One argument sometimes advanced is that the intense heat generated by the huge fires destroys the valuable humus content in the soil beneath, but it should be borne in mind that the main body of heat rises and investigations have shown that the destruction of humus from this cause is much less than was once thought. It must be conceded, however, that there is destruction of the uppermost layer of humus and that repeated burning at fairly short intervals can lead to noticeable soil degradation. The value of the operation has also been questioned by Gourou (1968, 32–3) who estimates that following the clearing of a patch of forest between 620 and 1130 tonnes of organic matter can be destroyed per ha by such burning, and he argues that this material could be more profitably employed in the form of timber, firewood, wood-pulp, leaf manure and products obtained from the distillation of wood. Between 110 and 165 kg/ha of nitrogen are lost to the atmosphere during the burning. Destruction of this kind is increased by groups such as the Bemba and Lala from Zambia who strip and clear an area six or eight times the size of the patch which is actually to be cultivated in order to increase the amount of fertilising ash from the burn.

This of course is one argument in favour of burning, that it produces considerable amounts of fertilising ash which enrich the soil in phosphorus and potash, though it is likely that the enrichment lasts only for a year owing to the excessive leaching which is normal in the humid tropics. Another point is that the burning destroys weeds and their seeds, though after the first year seeds blown in or carried in by natural fauna from the neighbouring forest germinate and in subsequent years weed infestation can become unmanageable; new land must then be sought elsewhere. Clark and Haswell (1970, 53–4) argue that there is a crying need for herbicides though these need careful application if they are not to kill the crops as well as the weeds. It should, however, be borne in mind that weed infestation often goes hand-in-hand with soil degradation and to the extent that this is true in particular cases herbicides would not help.

Immediately following the first rains the sowing and planting of crops begins. Several crops are normally planted on the same plot so that the farm yields not only a main crop but also secondary crops which vary the diet. Thus, for example, in southern Nigeria Hailey (1938, 888) states that of a selected number of yam plots chosen for observation 99 per cent grew pumpkins as well as yams, 93 per cent grew maize, 92 per cent groundnuts, 90 per cent red pepper, 80 per cent beans and 24 per cent cassava. Similarly the Iban of Sarawak grow a wide range of subsidiary crops on the land which is sown with upland rice. They grow ensabi (a kind of mustard plant of which the leaves are eaten), cucumbers, pumpkins, luffa and gourds, all of which are ready to eat before the rice is ready for harvesting, while near the edge of the clearing and close to the farm huts or watch-shelters are grown other plants such as cassava, maize and pineapples. Over parts of West Africa "in-farms" form a notable feature. These are small patches of land adjoining compounds within villages on which are grown crops such as maize, cassava, vegetables and even cotton. They are comparatively well fertilised with refuse and they play an important part in helping to diversify a generally monotonous diet.

The actual farming calendar and the crops grown in any given locality depend partly upon the climate, especially upon the length of the rainy season, and partly upon traditional practices. These points are illustrated in the case of West Africa in Figs. 15 and 16, the first of which shows the pattern of farming throughout the year in the Gambia and Sierra Leone; the former territory lies in the northern Grain–Legume Belt and the latter in the Rice Economy Region (Fig. 16).

While Fig. 15 shows how the pattern of farming changes throughout the year, Fig. 16 shows the main spatial variations in crop growing over West Africa. The main subsistence "upland"* crops grown in the Grain–Legume Economy Belt which lies immediately to the south of the Sahara and of which the Gambia forms part include various forms of millet, Guinea corn, groundnuts, pulses and cassava while other crops such as maize and cotton are also grown. Swamp rice has considerably increased in importance in this region since the close of the Second World War. In the Rice Economy Region of which Sierra Leone forms part the more extended rainy season lengthens the farming year and permits a greater variety of crops. Swamp rice is of great importance especially on the coastal swamps and along river valleys but the region has long been noted for the widespread growing of upland rice which gives the region its name. In some parts of the region cattle are pastured on the dried-out floodlands between January and May and crops such as maize, cassava, sweet potatoes, yams, cocoyams and various vegetables have been introduced on the higher floodlands which dry out first after the rains slacken. These crops are harvested from April onwards. On the uplands maize, cassava, and sweet potatoes are common while other crops include groundnuts and cotton which are in part grown for sale. In the Grain–Legume Belt the only cash crop of widespread importance is the groundnut, though cotton is grown in some areas (in parts of Northern Nigeria and on the Inland Delta of the Niger in Mali, for instance), but in the rice region a much wider variety of cash crops is produced including palm products, kola, coffee, cocoa, rubber and bananas. The tree crops are not, of course, produced under the shifting cultivation system.

* In West Africa the term "upland" means simply land above flood level. It does not necessarily denote land of any considerable elevation.

AGRICULTURE—SHIFTING CULTIVATION

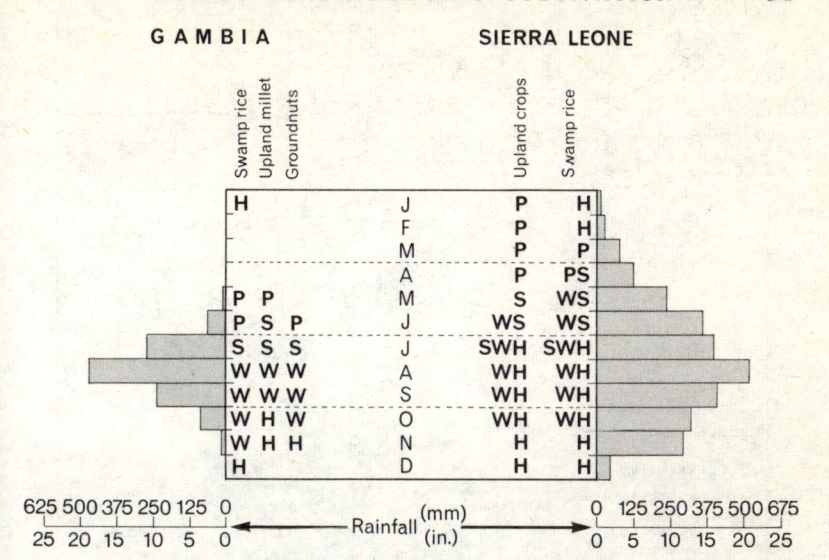

GAMBIA   SIERRA LEONE

Swamp rice	Upland millet	Groundnuts		Upland crops	Swamp rice
			J	P	H
			F	P	H
			M	P	P
			A	P	PS
P		P	M	S	WS
P	S	P	J	WS	WS
S	S	S	J	SWH	SWH
W	W	W	A	WH	WH
W	W	W	S	WH	WH
W	H	W	O	WH	WH
W	H	H	N	H	H
H			D	H	H

625 500 375 250 125 0   (mm)   0 125 250 375 500 675
25 20 15 10 5 0 ← Rainfall (in.) → 0 5 10 15 20 25

FIG. 15.—Farming calendars in the humid tropics.
P: preparing farms; S: sowing; W: weeding and minding growing crops; H: harvesting. The rainfall graphs are based on figures for Banjul (Gambia) and Njala (Sierra Leone).

The other two main regions shown on Fig. 16 are almost self-explanatory. Root crops, especially cocoyams, yams and cassava, have traditionally taken the place of upland rice as the main staples in the south roughly midway from west to east in the Ivory Coast, though various varieties of upland and swamp rice are important locally. Maize is the only cereal of widespread importance in this Root Economy Region though the oil palm grows almost everywhere; vegetables and fruits are also widely grown but vary in importance from place to place. As in the rice region tree crops form the leading cash crops. The Grain–Root Economy Belt, as the name perhaps suggests, is transitional between the Root Economy Belt to the south and the Grain–Legume Belt to the north.

The annual harvesting of the more quickly-growing crops begins fairly early during the rains and that of later crops continues for some time after the rains have finished. There may or may not, according to climatic and local circumstances, be a gap before work recommences for the following farming year. These points are all illustrated on Fig. 15.

We should now turn our attention to the hotly debated question regarding the whole status of shifting cultivation, for some authorities condemn it as a wasteful and harmful system while others take a

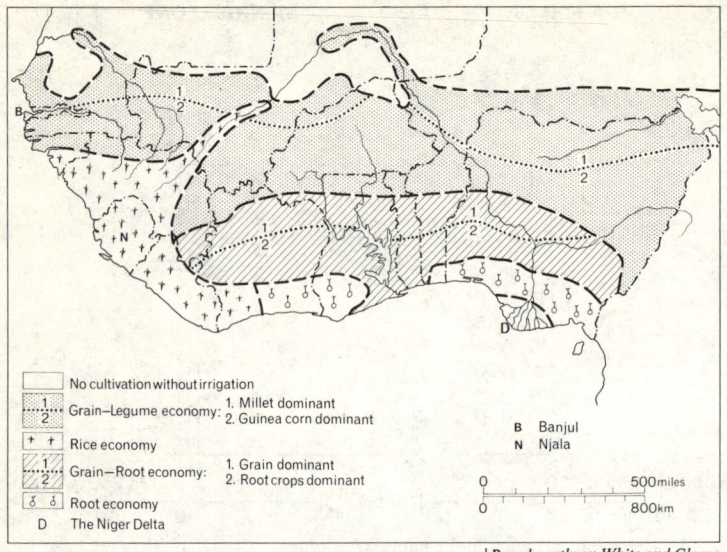

Fig. 16.—West Africa: types of subsistence economy.

contrary view. There is in fact some preliminary debate as to the exact nature of the system which, as might reasonably be expected, shows considerable variations in technique from area to area. Some writers, for example, distinguish between true shifting cultivation and rotational bush fallowing. In the former the clearing of plots proceeds according to the needs of the moment; there is no fixed pattern of clearance, and when further bush clearing is no longer practicable in a particular locality the whole village moves elsewhere. Clearance then proceeds, starting with land nearest the new village and steadily moving outwards until the cultivated plots are so far from the village that excessive amounts of time are taken up simply in movement between the settlement and the farms. It is then time to move once again.

Under the bush fallowing system, however, settlements are more permanent and each village owns the land around it. Boundaries with lands of neighbouring villages are fairly clearly defined. Bush clearance over the village land proceeds year by year according to a pattern so that after a given number of years all the available farming area has been cropped and the cycle recommences. The number of years which is taken up by a single cycle of cultivation depends upon the total area concerned and the population of the village. Other things being equal a larger community will need to cultivate a larger area each year in order to provide itself with the food which it needs and this may result in a speeding up of the cycle; in other words, any given plot will be cultivated more frequently.

It is possible to recognise certain features which are characteristic of any form of shifting cultivation and for the purpose of this particular study we shall take the following: shifting cultivation includes those types of agriculture which do not include permanent field cultivation as a main element (in a sense it is the fields which are rotated rather than the crops as the same crops are grown each year but on different cultivated plots while the old plots are abandoned to fallow); bush clearance is accomplished by slash and burn methods; short periods of cropping alternate with longer fallow periods; and no use is made of animal power. Only human labour is used in the various farming processes, a point which receives further attention later.

It must be admitted right away that there is much to be said in favour of the system which many writers see as the only possible response of non-technological societies to soils of generally limited fertility. Some argue in fact that shifting cultivation in one form or another was probably the earliest developed farming technique of pioneer agriculturalists in the world, and the system, on this basis, may be regarded as a stage through which most agricultural communities have passed at some time. It may perhaps best be viewed as a temporary though essential stage which continues until it is outmoded by more productive techniques. It may be possible, for instance, to perceive without too much difficulty links between shifting cultivation of the bush fallowing type and the medieval three-field rotational system of the England of years past.

One main point in favour of shifting cultivation is, of course, the fact that it makes possible cultivation of soils of inferior quality, soils which cannot sustain crop cultivation every year without the enrichment which is beyond the power of traditional societies to provide. It is important to remember that soils in many parts of the world which today are considered fertile were frequently of mediocre quality before they were scientifically managed; many British soils, for example, come under this category. In the absence of such practices as continued fertilisation, crop rotation and mixed farming, to mention some of the techniques employed to maintain and even increase soil fertility, it is only the abandoning of cultivated land to fallow which enables it slowly to recover its limited fertility. It may be argued, therefore, that shifting cultivation is a necessary response to inherently poor soils but that, as soil management techniques develop, it is likely to be replaced by more progressive husbandry.

Among its other advantages there is no doubt that the system also provides a method of dealing with the problem of weed infestation, which is such a severe problem that without the burning of the freshly-made clearings crops would probably be smothered beneath a blanket of weeds. It is also true that the slash and burn method is the only practicable one open to primitive cultivators who have simply tools of indifferent efficiency. It is possible to argue that the system of bush clearance is destructive of forest growth, and this can certainly be true if the land is overused, a point which was discussed in Chapter I and which is examined later, though it has been suggested that in general the extent of the destruction may not be as great as is sometimes supposed. Some authorities believe that the destruction of natural vegetation in tropical rain forest is not easily achieved and it is true that normally only any undergrowth which may be present and the smaller trees are removed, larger trees simply being lopped. There is, however, some doubt as to whether this argument is of general application as we shall see later. It is, of course, true as we have previously observed that burning produces a fertilising ash, but this is likely to be only advantageous in the short run.

Nevertheless, it does seem to be true that local communities frequently prefer the shifting cultivation system and examples are not lacking of groups who have adopted it even after formerly practising sedentary farming. Hodder (1973, 100) mentions the instance of the Cabrais of northern Togo, some of whom have moved southwards from their overpopulated homelands to settle in parts of the comparatively empty Middle Belt of the territory. As this happens the farmers concerned rapidly forsake the comparatively sophisticated farming methods which they have followed for centuries in their homeland and adopt shifting cultivation techniques. It is sometimes argued that this technical change demonstrates a preference on the part of peasant farmers for leisure as less labour is required to produce any given amount of food using shifting cultivation methods than is the case under sedentary agriculture. Geddes (1954) has pointed out that the Land Dyaks of Sarawak prefer to grow upland rice under shifting cultivation methods rather than swamp rice, though acre for acre the swamp yields are higher (1903 kg/ha (1700 lb/acre) in the swamps as against 1578 kg/ha (1400 lb/acre) on the uplands). The main point at issue, and one well understood as a practical matter by the Land Dyaks, is that while in the swamps the rice yield per man-hour of labour is 0·87 kg, working at a labour input of 2165 hours ha, on the uplands it is 0·95 kg (labour input 1663 hours ha). When land is comparatively abundant yields per unit of labour are clearly of more importance to cultivators than yields per unit area.

While all these points may be fully admitted, however, it still remains true that the shifting cultivation system continues to evoke much criticism. We have already noted, for example, some doubts which have been voiced regarding the burning method of bush clearance, but very important other questions must be raised. The following are probably the most important.

1. There seems to be little doubt that overuse of a region for shifting cultivation does result in vegetational degradation as the original forest is given no opportunity to regenerate itself. Instances of this actually happening are not lacking. Freeman (1955), for example, draws attention to the situation among the Ibans of Sarawak who grow upland rice using a *ladang* system. In the Iban area a single harvest followed by a long fallow does not degrade either the soil or the vegetation owing to their powers of regeneration but if crops are harvested in successive years or if the period of fallow is reduced below the absolute minimum of 4 years the vegetation suffers. If this does happen the forest is no longer able to re-occupy the worst-affected clearings for they become infested with tall grasses which grow from rhizomes. Near the end of the dry season these grasses are reduced to highly inflammable straw which is easily ignited from neighbouring fires lit to make other clearings. The subsequent burning of the grass devours other growing plants but the rhizomes beneath the surface of the ground remain active. So the grasses maintain their hold though the forest would eventually return if the burnings ceased.

Another example of this type of destruction was given earlier (p. 89) when it was shown how some groups strip and clear areas greatly exceeding in size the areas of the patches actually to be cultivated in order to secure additional supplies of potash from the burning. Ng (1971) states that in Thailand some 3m ha, representing almost 70 per cent of the evergreen forests which have taken thousands of years to reach their present climax stages, have been denuded by shifting cultivators and that a further 40,000 ha of primary and secondary forest are being cleared each year. Further disadvantages also incurred include the destruction, largely through burning, of potential timber supplies to the value of £250,000 annually while the stripping of forests from watershed areas which is proceeding at some speed will in time affect rice growing on far away lowlands. This is because the successful cultivation of swamp rice on the alluvial plains of southern Thailand depends ultimately upon the conservation of water supplies in the remote northern parts of the country. Denudation of forests in watershed regions will lead to worsening floods during the rains but river levels will more rapidly fall as the rains slacken and close. There is clearly a connection here with a point made in Chapter I that continued forest clearance in the humid tropics is likely to lead to the development of savanna, a vegetational type which is of such widespread occurrence today yet which is not a natural climax but one induced by man.

2. The point has been made in the last chapter that shifting cultivation encourages communal rather than private land ownership. It could hardly be otherwise as new clearings are made by communal labour year after year as the old ones are abandoned. In many instances a cultivator does not lose all his rights over a patch when he has abandoned it for he still retains proprietorial rights over fruit trees which he may have planted. This right is particularly important in parts of Africa in connection with the oil palms which frequently appear to be growing in unused forest but which invariably have been planted during a previous period of cultivation and of which the produce belongs to the planter or his family. It is also true that a cultivator often exercises superior rights in succeeding years to any patch which he may himself have cleared and cultivated.*

Such rights as these, however, valuable as they may be, are by no means the same thing as direct ownership of land and experience elsewhere has shown that such ownership is a necessary prelude to improved farming efficiency. It was the enclosures, painful as they were, which helped to make possible the English Agricultural Revolution, while it was the Mexican revolution of 1910 which paved the way for the transformation of a semi-feudal economy, largely through the distribution of land to peasant farmers. It is, however, only fair to say that this movement did not gain much impetus until the government of President Cardenas (1934–40) when the land was assigned to the people, and this action initiated a period of rapid economic growth which could never have taken place under the *ancien régime*. Although there are major differences between these two examples the lesson is clear: agricultural development is normally delayed until the cultivators themselves have a strong stake in the enterprise, and the strongest stake of all is the ownership of the land. Under the present communal system still dominant in the humid tropics the entire weight of tradition and public opinion is against the innovator and the individualist, precisely those who would

* This general point is examined in more detail in Chapter VI, pp. 141–2 below.

be in the van of economic progress, while there is in any case no incentive to try to improve land which is not the property of the cultivator and which will be abandoned in due course. There is therefore a strong case for arguing that the system of shifting cultivation is self-perpetuating and a very real barrier to economic and social development.

3. A disadvantage which accompanies true shifting cultivation, though not its bush fallow derivative, is the periodic movement of settlements for this means that it is extremely difficult to provide public services to village communities. Schools and hospitals, for example, and indeed any forms of public services, necessarily need permanent bases. The provision of piped water supplies and modern hygienic methods of sewage disposal, to take other obvious instances, are quite incompatible with the periodically moving settlement. Where settlement is dispersed a similar difficulty arises. In parts of Central America, for example, many so-called "towns" have little in common with the towns of the western world. The Central American town is frequently not primarily a place of residence but a community centre or meeting place for the area it serves and only public servants such as administrators and priests actually live there. A similar situation arises in connection with the periodic market which is frequently held at a convenient meeting place, but this meeting place is not necessarily a centre of any permanent settlement; this phenomenon is examined by Hodder and Lee (1974, 136–47) and also in Chapter VIII below. Such a marketing pattern is frequently typical of a shifting cultivation community, with its attendant disadvantages of a lack of storage facilities, rudimentary transport services, and a population density too low to support the permanent trading bases which are normally signs of economic growth. The fundamental point is that when community living is so disparate it is a matter of extreme difficulty to initiate any form of social or economic progress, and it is unlikely in the extreme that such progress will take root and develop of its own accord under such conditions.

4. There seems little doubt that the *ladang* system is inefficient in its use of labour, a point which does not contradict what was said earlier (p. 94) when we were comparing the labour-output ratios of two different systems of production.* In fact the shifting culti-

---

* Inefficient, that is, from the point of view of returns to labour. It is possible with the help of mechanisation greatly to increase the output of rice fields per unit of labour and this is, in fact, done in the rice-growing areas of the southern U.S.A., but the output per acre under these conditions is lower than under labour-intensive systems such as those of south-east Asia.

vator seems frequently to be caught between two irreconcilable disabilities—shortage of labour on the one hand and surplus labour on the other.

Clark and Haswell (1970, 93 *ff.*) refer to the widespread belief that L.D.C.s such as those of the humid tropics normally have a considerable surplus of agricultural population and this belief is often accompanied by the notion that this surplus labour could be usefully employed in some other form of economic activity—perhaps in industry, for example. This is not infrequently advanced as one argument in favour of industrialisation in the L.D.C.s. Hodder (1973, 87) makes a similar point when he suggests that the term "population pressure" is too often loosely used whenever there exists a comparatively high density of population in any area together with accompanying poverty and need. There are good reasons for disputing these approaches. The concept of any "pressure" or "surplus" as applied to population or labour under these circumstances is at best a highly questionable one for it is to say the least highly unlikely that if the population of an area such as that which we are considering were significantly reduced there would be anything like a proportionate increase in living standards. It is also true that labour requirements vary tremendously from one part of the year to another. The variation is probably greatest in areas of markedly seasonal rainfall where the bush clearing, sowing and planting have to be telescoped into a very short period and where the harvest too makes great demands upon the available labour resources. At such times labour, instead of being a surplus commodity, "suddenly becomes the critically scarce factor of production" (Clark and Haswell). The situation is made far worse if, for instance, the onset of the rains is delayed (p. 28 above) or if there is illness. Gourou argues that it is a disaster to be ill during the planting season because this can slow down agricultural work sometimes with catastrophic results, and this argument is supported by other writers. Akehurst and Sreedharan, for instance, writing in the *East African Agricultural and Forestry Journal* (January 1965) have shown that a month's delay in planting maize in Tanzania can result in harvest losses of between 15 and 80 per cent according to the season, while experiments conducted upon cotton growing in West Africa have demonstrated that an average loss of production of about 8 per cent can be expected for each week's delay in planting.

It is easy to see why in the face of such evidence the question of labour surplus is one very much open to debate, and in fact in many areas the factor limiting production is shortage of labour. The situation is by no means as simple as it sometimes appears for it is equally true that there are certainly periods in the year when labour is grossly underemployed and one is led to ask whether workers could not be productively employed at such times.

The answer seems to be that *within the context of shifting cultivation* this is probably not possible unless the dry season is fairly long and alternative employment can be sought, for example in swamp cultivation. We must bear in mind that we are dealing with a situation in which any amount of land above the norm can be cleared only with a very considerable additional expenditure of effort while the clearance and use of larger plots year after year would mean more frequent cultivation of the village lands with consequent diminution of yield on soils which may be subject to overuse and progressive impoverishment. In other words, beyond a certain labour input returns may be expected to move sharply downwards and this is, in fact, what seems to happen. In a very thorough study based on field work Haswell (1953) shows that in a Gambian village the relationship between man-hours worked and productivity is almost horizontal (the marginal product is about the same as the average product) as far as a labour input of about 700 man-hours per hectare but beyond that level the marginal product falls away rapidly. The interesting thing is that cultivators do in practice stop putting further labour into cultivation at about that level.

The point is clear; there are very definite upper limits to productivity inherent in the shifting cultivation system and additional use of existing labour will not change this fact significantly. The only way out of the dilemma is to change the system, and this is a point which we shall carefully examine later.

5. The question of population density becomes a very important one in the face of evidence such as that just cited. It has long been recognised that the shifting cultivation system cannot of itself support a dense population and that beyond a certain level the variant known as bush fallowing develops. Even under that system, however, there is still a very real limit to the numbers of cultivators and their families which can be supported. Gourou (1968, 45) believes that with a 25-year rotation cycle (1 year of cultivation and 24 years of fallow) a population density of about 12 persons per km^2 (30 persons per sq. mile) may well be a likely maximum, though circumstances will vary from locality to locality. He gives, for instance, corresponding figures for parts of Rhodesia of 2·2 per km^2 (7·7 persons per sq. mile) and for Sumatra of 11·5 per km^2 (30 persons per sq. mile).

There seems to be little doubt that shifting cultivation cannot maintain large populations nor can it support significant population increases. The inevitable reaction to overpopulation is the shortening of the fallow period in a desperate attempt to produce an adequate food supply. This has happened, for example, in Sierra Leone where in some areas the fallow is as short as 6 years, so critical is the situation, while in parts of south-eastern Nigeria in Iboland any given plot is likely to be under cultivation every 4 or 5 years. The inevitable result of this kind of situation is a deterioration in the soil which in extreme cases may be ruined completely; this in fact has happened in some parts of Sierra Leone where now useless areas with a rock-hard surface of lateritic pan (*carapace latéritique*) are by no means uncommon.

A CASE STUDY: THE MAYA

The whole question of the connection between shifting cultivation, population and culture is a fascinating one, and one possible example has been examined by Cook (1909) who has put forward a striking and very important hypothesis regarding the downfall of the Mayan empires, the first one of which flourished in Central America between the second and eighth centuries A.D. It then quite quickly declined and the last of the original Mayan cities were deserted in the early part of the tenth century A.D. Coincidentally with this rapid decline came the foundation of a new empire and new cities were founded to the north and to the south of the former empire (Fig. 17); Gourou (1968, 56) says that it appears as if the Maya left their original home over a period of about a century and settled to the north and south of it. After this diaspora the Maya staged a remarkable revival especially in Yucatan and a new empire was established during the tenth century as the magnificent ruins of Chichen Itza and other cities demonstrate. This Second Empire lasted until the fifteenth or the sixteenth centuries but it seems already to have reached an advanced state of decay when the Spaniards arrived in 1527 and finally destroyed it.

It is important to note that the Mayan civilisation reached very considerable cultural heights; it was easily the most advanced in the whole of the New World before the advent of the European. The beauty and the style of the Mayan buildings are remarkable; as early as the third century the Maya were building enormous temples and pyramids of stone and mortar adorned with elegant sculptures and wall paintings. They evolved a pictographic form of writing and an accurate calendar and it is clear that they had arrived at a considerable understanding of

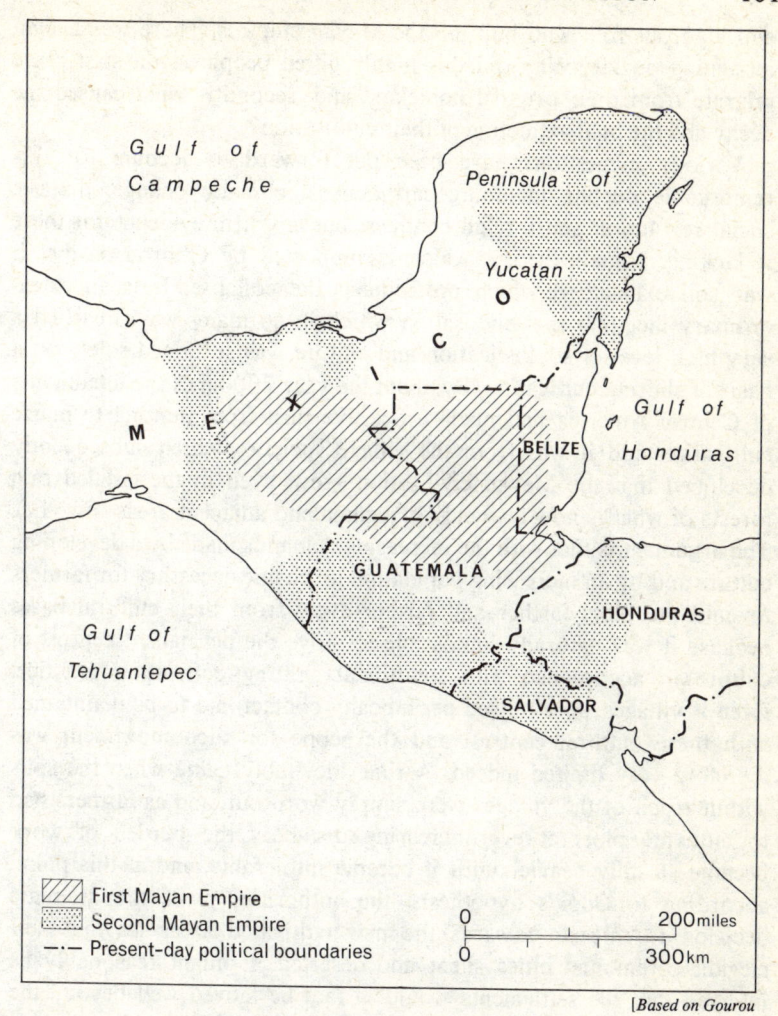

FIG. 17.—The Mayan Empires.

astronomy and mathematics. From house counts on the spot it has been estimated that at least in parts of the First Empire the population density reached 180 per km² (470 persons per sq. mile)—a very high figure for a region which today is covered in tropical rain forest except in some savanna areas and which is almost uninhabited. The few inhabitants who at present live in these parts of Mexico, Guatemala, Salvador and Honduras are disease-ridden and pitiful and they must be counted among the poorest and most underprivileged inhabitants of the humid tropics; they have nothing in common with the highly gifted and vigorous folk who built up the Mayan empires. There we have an intriguing puzzle. Why did this highly gifted people in the first place migrate from their original homeland and, secondly, what caused the decay and the final extinction of their civilisation?

Various suggestions have been put forward to account for this remarkable decline including earthquakes, climatic change, disease, social revolution and foreign conquest but any firm evidence for these is lacking. Cook's theory, which is supported by Gourou, is that it was soil exhaustion which precipitated the collapse. It is an extraordinary fact that this civilisation which in so many ways reached a very high level of sophistication and culture was sustained solely on a basis of shifting cultivation. By about the year 2000 B.C. the inhabitants of Central America had made a decisive turn from hunting to maize cultivation and it was from the village societies which subsequently developed that the Mayan civilisation grew, even in the tangled rain forests of what is now lowland Guatemala and adjacent areas (Fig. 17). The argument is that with the increasing demands made by a developing culture and by an increasing population it became necessary for farmers to cultivate plots farther and farther away from their cultural bases because it was naturally impossible to move the permanent centres of culture in accordance with traditional shifting cultivation practice. Even if villages were moved periodically contact had to be maintained with these cultural centres, and the scope for such movement was therefore very limited indeed. A time inevitably came when the soils within reach of the villages were simply worn out, and as farmers had to cultivate plots at ever-increasing distances the burden of work became steadily heavier until it became intolerable, and at this point, according to Cook's hypothesis, the authorities made the dramatic decision to migrate towards the new cultivated areas leaving their elegant ceremonial cities silent and deserted. It might reasonably be inferred that the settlements would in fact be moved well beyond the areas actually cultivated in search of virgin soils and forest.

Such a development would explain the centrifugal movements of the Maya outwards from their original homeland (Fig. 17) and it might also explain why these movements continued over a period of about a century for it is unlikely that all the original centres would feel it necessary to move at the same time. It might even explain why the Maya began to place more emphasis upon human sacrifice in the Second Empire for they might well have felt a more desperate need to appease the gods and at the same time to restrict population growth; there is also the possibility that such practices, unknown in the First Empire, may have been brought in by Toltec invaders from Mexico. Whatever the causes of these developments there is no doubt of the results; we know, for instance, that in times of famine living victims, preferably children, were thrown into the 40-m deep "Well of Sacrifice" at Chichen Itza in an attempt to appease the rain gods.

The final chapter in this extraordinary story is probably similar in most respects to that which records the decline of the First Empire, and it tells of ever increasing difficulty in providing food from the *milpa* cultivation system and of a corresponding decay in the whole social fabric. This hypothesis is supported by the fact that even after the lengthy period which has elapsed since the final collapse of the Mayas the forests in the lands of the two empires are still secondary growths, suggesting that the original forests were so thoroughly cleared and the soils were so degraded that new forests have had the greatest possible difficulty in establishing themselves. Indeed, quite extensive areas are still covered with savanna. It seems that the whole culture became weaker until it collapsed completely before the onset of the Spanish invaders of the sixteenth century, never to recover.

FINAL CONSIDERATIONS

It has been thought worth while devoting some space to this striking example of cultural development in the humid tropics as it highlights some extremely important lessons.

1. It should dispose of any lingering notion that the inhabitants of the humid tropics are inherently incapable of human achievements comparable to those which have taken place in other parts of the world. There were elements in the Mayan civilisation which will bear comparison with any similar achievements elsewhere.

2. It sharply demonstrates the long-term deficiencies of the shifting cultivation system. Here is a case of an ingenious people who successfully developed a remarkable and virile culture which managed to survive even a wholesale uprooting and transplanting. No doubt many factors contributed to the demise of this civilisation, and this is not the place to talk of the human factors which were almost certainly involved, but the overriding probability seems to be that the primitive *milpa* system could not continue to support this civilisation indefinitely. The wonder is that it succeeded in doing this for so long.

3. The inescapable corollary to 2 above is a confirmation of a point which has been made earlier in this chapter, that any advanced form of civilisation in the humid tropics must modify and change

its economic basis from shifting cultivation to a more productive system. It is perhaps not entirely a useless exercise to wonder what further heights the Mayan civilisation might have scaled if its leaders had directed into technological channels at least part of the energy which was taken up by the maintenance and development of their particular kind of theocracy. No one should underestimate the strength of the essential contributions to the growth and maintenance of any form of civilisation which come from the human mind and spirit, but at the same time it seems to be the case, whether we like the idea or not, that one essential basis for human development and achievement lies in technological progress, and this is just as true in the humid tropics as it is elsewhere in the world.

Chapter V

Agriculture in the Humid Tropics: Other Forms of Farming Activities

No other system of cultivation in the humid tropics is as widespread as shifting cultivation, but this is not to say that no other systems exist: indeed, sometimes other cultivation techniques are locally more important than shifting cultivation while they certainly are of widespread occurrence. We should at this point turn to an examination of the other more important farming patterns.

WET-LAND CULTIVATION

In many parts of the humid tropics wet-land farming is carried on with great vigour and success. The crop produced under these conditions is, of course, swamp rice and it is south-eastern Asia which has led the way in this type of agriculture. It is not surprising, in view of the great success of this farming technique, that many authorities urge its expansion throughout the tropical world where conditions permit, and many governments are indeed actively encouraging the cultivation of swamp rice.

It is very easy to point to the advantages of wet-land cultivation. A rice crop provides the highest nutritive value per unit area of all tropical cereals while soil fertility is not the vital factor which it frequently is in the growing of upland crops; the important requirements are warmth and moisture. It is frequently possible to secure two harvests a year and it is an important point that the effects of drought are least felt on flooded riverine or coastal lands. Furthermore, rice cultivation does not give rise to soil erosion.

On the other hand there are very real difficulties which the rice cultivator has to face. Some areas, for instance, must be classed as unsuitable for wet-land farming for reasons of topography, for flood plains may not exist or if they do they may be of very limited extent. Where suitable land does exist the labour input required initially to prepare it for rice cultivation is very high indeed, for the clearing of swamp vegetation is tremendously arduous and unpleasant work while the low bunds which separate the fields must also be constructed and maintained. Sometimes it is possible to terrace sloping ground for wet rice production but this, too, is very strenuous work. While such terracing is sometimes possible, as has been strikingly demonstrated in the Philippines, there are certain requirements which must be met if this is to be successfully achieved. Labour, for instance, must be plentiful while rainfall must be heavy during the growing season as water in high-level terraces cannot be replenished following losses through seepage and evaporation except through this means.

Environmental requirements for successful wet-land cultivation are in fact considerably more exacting than is frequently realised, for as we have just seen, not only must the topography be favourable, but also the climate. Very large amounts of water are essential to provide the necessary flooding for example, and this normally means that rainfall must be heavy though it is true that some riverine lands in comparatively dry areas can be inundated by flood waters received from areas of heavy rainfall upstream. The inland delta of the Niger in Mali is a case in point. Within the humid tropics annual variations in rainfall can markedly affect crop yields as they affect the amount of land flooded year by year as well as the duration of the floods. Quite lengthy dry, sunny periods are needed for the ripening and harvesting of the crop and it is for this reason that truly equatorial climates with their constant cloud cover and lack of a dry season are not well suited to rice cultivation.

One other obstacle to the extension of rice cultivation throughout the humid tropics is cultural, particularly in the American and African tropics where the production of swamp rice is not an indigenous activity. In these areas the staple foods locally produced include root crops (yams, cassava, cocoyams, groundnuts and sweet potatoes, for instance) and upland cereals, particularly upland rice, maize, various kinds of millet and sorghums, and these are the foods to which the inhabitants are accustomed. It is not always realised that innate resistance to change, including changes in long-established eating habits, can be extremely powerful in traditional societies, and in many regions it would require a major change in eating habits, as well as in cultivation techniques, to initiate a move towards the establishment of rice as a staple article of diet.

Even so, although we must recognise that the extension of wet-land cultivation throughout the humid tropics is not practical, at least in the short run, it is still true that this form of agriculture is increasing in importance, for instance on the Amazon lowlands, on the *fadamas* (seasonally flooded riverine lands) of northern Nigeria, and along the coastal swamps of Sierra Leone, and it may well continue to do so. The example of south-eastern Asia reminds us of the potential of rice growing for not only can it sustain dense populations but in many instances the communities concerned have reached high cultural levels. It is indeed the case that the regions within the humid tropics which can boast of the richest cultural traditions are those parts of Monsoon Asia which are based upon low-lying flood plains where century after century flooded rice fields have produced regular and dependable food supplies. While flood plains of comparable extent and productivity may not exist elsewhere in the humid tropics irrigable plains are by no means uncommon as the examples given above show, and it seems likely that an increasing awareness of their potential will in time lead to their greater use.

CASH CROP FARMING

So far we have been considering for the most part subsistence farming in the humid tropics but it would be misleading to suggest even by implication that cash crop farming does not exist. This type of farming has in fact developed in many areas and it would be very surprising if it had not in view of the powerful trading links which have been forged between the industrial nations on the one hand and much of the humid tropics on the other. Cash crop farming was at first encouraged to meet the demands of the newly developing countries, as Britain and the other industrialised territories then were, for foods and vegetable raw materials which cannot be produced in temperate lands. There is a sense in which it is true that colonialism sprang from economic roots as the realisation of the value of colonies as producers dawned upon the developing countries of the time, particularly those of North-West Europe. It is easy today to underestimate the strength of the economic impact which early colonialism made upon the humid tropics; it was, perhaps surprisingly, as long ago as 1926 that McPhee could write a text-book entitled *The Economic Revolution in British West Africa* and the recollection of this may serve as a salutary corrective to those who appear to believe that in the economic field colonial affairs stagnated completely until the close of the Second World War.

The results of this interaction of supply and demand between the humid tropics on the one hand and the developing lands of temperate latitudes on the other varied very greatly. Sometimes, as in the case of Ghana, they greatly benefited the tropical territory concerned. The comparative prosperity brought by the cocoa farmers of the south very greatly strengthened the economic base of the territory and it was largely because of the resulting economic strength that the Gold Coast was able to lead the way in the march towards independence. Sometimes, however, the territory suffered; the case of the savage mishandling of the rubber producing industry in Brazil has already been referred to (p. 38) while it has also been argued that various monocultural agricultural enterprises in the same country have not been helpful to long-term development (*see* p. 116 below).

It is important to bear in mind that as cash crop farming is introduced to an area it does not normally displace subsistence agriculture. Not only do the two forms of production normally exist side by side but it is frequently impossible to draw any clear dividing line between them. It seems likely that purely subsistence agriculture hardly exists at the present time, but at the other end of the scale it is also unlikely that purely cash crop farming exists either, except in the extreme case of the plantation. Most tropical farmers continue to produce subsistence crops but grafted on to this activity is an element of cash cropping. The actual amount of crops grown for sale is likely to vary from year to year in accordance with such factors as crop yields and the strength of market demand which will manifest itself in price variations. It is thus difficult, indeed it is normally impossible, to separate the subsistence farmer from the cash cropper though it may be reasonable to suggest that if a farmer simply sells any surplus produce or if he merely sells under market persuasion (*e.g.* if prices are unusually high or if he wants extra cash to make selected purchases) he should be viewed as a subsistence producer. If, on the other hand, he runs his farm with the deliberate intention of selling most of his produce he should be classed as a cash crop farmer.

Even the argument in the foregoing paragraph may suggest a precision which frequently does not exist, for many farmers undoubtedly will lie close to the margin between subsistence and cash cropping and their market intentions in any one year will be uncertain until after the harvest. It is only when such a farmer knows just how much produce he has at his disposal and can assess its value in terms of prevailing market conditions that he will decide what proportion of his harvest he will sell and what he will keep for his own use. And even the apparently simple term "sell" is not always a precise one. There can be a real difference between, on the one hand, selling substantial amounts of farm produce for cash prior to the despatching of the produce to distant markets and, on the other hand, the exchanging of comparatively small amounts of the same produce locally for other goods. Such local exchange may be little removed from barter.

AGRICULTURE—FORMS OF FARMING ACTIVITIES

Many actual examples can be given to show the working of this pattern in practice. Firth (1946, 22), for instance, shows how a Malay farmer with the help of his wife grows enough rice each year for himself and his family. This gives him his basic food requirements. After the rice harvest he grows vegetables partly for subsistence and partly for sale in the local market, while in addition he is likely to own a small plot of rubber trees, the produce of which he sells to a Chinese dealer for future export. This "spreading" of economic activity gives the farmer a security which he could not hope to enjoy if he specialised on cash crop production for such specialisation would leave him at the mercy of market forces which are quite outside his control. Phillips (1964, 2–4) describes a comparable farm on the island of Luzon. Each year the farmer and his family plant maize, sorghum and upland rice on about 0·24 ha (0·6 acres) of farm land. A plot is cultivated for between 2 and 10 years after which it is abandoned and a new one is cleared. These crops produced under shifting cultivation methods are used primarily as food for the family but small amounts may be sold locally from time to time. There is no set pattern in this. Also on the farm, however, are about 2·4 ha (6 acres) of coconut trees which are planted partly on narrow strips of flat land along a valley floor and partly on the more accessible valley slopes. The trees are planted in rows whenever possible, otherwise according to the dictates of relief. The coconuts are harvested in June, July and August while they are still green enough to cut open easily. The copra is extracted from the nuts and after drying it is carried to buying stations in nearby villages where it is bought by agents of copra companies.

In many parts of the humid tropics the transition from purely subsistence to subsistence with cash crop farming has been deliberately fostered by governments through the medium of taxation. Of the justification for the imposition of taxes upon a peasant community this is not the place to speak, but it is clear that if taxes are demanded money must be earned to pay them. In this connection Winter (1956) has outlined the stages of agricultural development which are likely to ensue and his often quoted scheme is of value in helping the student to recognise the various stages through which a developing agricultural community is likely to pass. Although Winter's argument was based upon the need of a community to raise money for taxes it is likely that a similar sequence would develop in any case as the community awakes to the desirability of securing monetary rewards which can be spent for the satisfaction of personal needs. The stages of development are as follows.

1. *Pure subsistence farming*, including hunting and collecting economies. No taxes are levied and there is no need of money. No cash crops are grown and there is no import or export of labour.

2. *Subsistence farming, but with taxes levied.* Some cash crops are therefore grown to secure money for paying the taxes while some workers may seek paid employment elsewhere, primarily for the same purpose.

3. *Subsistence farming and cash cropping.* The growing of cash crops forms an important part of the year's farming activity in this stage. The cash is used to purchase a variety of commodities and the payment of taxes now forms a comparatively small part of the yearly cash outlay. There is little, if any, export or import of labour.

4. *Cash cropping with subsistence farming.* This stage is reached when the desire for cash becomes even greater and many male workers may travel long distances to secure comparatively well-paid employment. A well-known example of such a class of workers is that of the "Strange Farmers" of the Gambia (Jarrett, 1949) and the *navétanes* of Senegal, migrant workers who move seasonally to help with the cultivation and harvesting of groundnuts. Many of them come from Mali.

5. *Agricultural wage economy.* Most workers are at this stage engaged in paid labour as farming becomes more commercialised, while others may work on plantations. This particular form of economic development is examined later in this chapter.

6. *Industrial economy.* With the growth of urban and industrial development more and more workers leave their villages for paid employment in town, mine and factory. This is a late stage of development and it is examined in a later chapter.

It will be clear from the foregoing that a growing demand for cash, whether voluntary or involuntary, is a very important motive in bringing about changes and modifications in traditional communities; in fact, it may well be the most important factor. At the present time, however, it can scarcely be claimed that cash crop farming plays as important a part as it should in the economic development of the humid tropics though there is little doubt that such development is essential for the eventual emergence of an urbanised industrial society. More will be said about this important point in a later chapter but we might at this point note that such development is not without its problems. The following are probably the more important issues involved.

1. An early result of successful cash cropping is likely to be an

increase in the *per capita* demands for land as cultivators are encouraged by the prospect of cash returns to spend more time on their farms and as the more efficient farming implements bought with cash enables them to cultivate more extensive areas. Such a development can lead to obvious difficulties in situations where land is relatively scarce and where there is danger of overcropping. This may indeed be a temporary phase which will disappear with increasing farming efficiency but it can pose serious problems in the early stages.

2. A related point is that as a result of the prosperity brought by cash cropping the farmers and their families will begin to eat more thus further increasing pressure on the land. Some of the additional food they will, of course, produce themselves but some they will buy from other producers. It is at this stage that the blurring which we have already commented upon sets in between subsistence and cash crop farming. Sometimes this increased demand may be met by bringing into production land hitherto unused such as inland swamps, but once again there may supervene a difficult transition period before new farming practices result in generally increased efficiency.

3. The economic independence and the success which often attends cash crop farming is bound at some time to encourage the farmer to attempt experiments to improve his land, but any such experiments are almost certain to run into difficulties as long as the shifting cultivation system is generally maintained. A conflict may thus develop between the old system of land tenure and use and the desire for permanent land occupancy and this must lead to stresses within the community.

4. The small-scale cultivator who ventures into the realms of national and international trade through the selling of crops will quickly find that he is dealing with forces over which he has no control and which he does not understand. Sometimes, during periods of rising prices, these forces will act to his advantage but at other times as world prices fall the reverse will be the case. This situation can be tremendously confusing to the unsophisticated peasant farmer who can experience considerable difficulty in deciding how much time and effort he should put into cash cropping, and this confusion can easily turn (or be turned) into an irrational anger against "those" who are believed to "manipulate" prices to their own advantage. We are here entering the sphere of politics and it is important to bear in mind the very close connection which exists between economic activity and politics. The case of cocoa production in parts of West Africa provides an outstanding example of this connection.

West Africa is easily the world's leading producer of cocoa and production lies in the hands of small-scale peasant farmers. The leading producer is Ghana and Fig. 18 shows fluctuations in the world price of high quality cocoa between 1950 and 1966. These fluctuations were very considerable and it is not difficult to discern possible links between cocoa production and prices on the one hand and political change on the other. It has earlier been suggested that it was the prosperity brought by the cultivation and export of cocoa which formed the economic basis for the emergence of the Gold Coast (as the territory then was) as the first colony to secure independence in 1957, though internal self-rule had been granted in 1951, and this fact alone provides a clear example of the political power of economic strength.

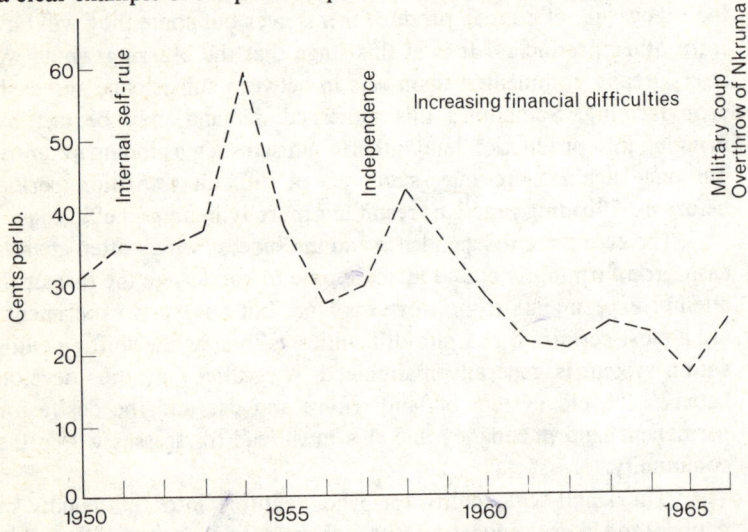

FIG. 18.—Cocoa prices and political changes in Ghana (1950–66).

The whole period following the close of the Second World War, however, was a very difficult one for cocoa producers because of the ravages of two diseases, swollen shoot and black pod, which sharply affected production and threatened the whole cocoa growing interests of the southern Gold Coast. The only known method of combating the diseases was the cutting out and burning of affected trees and an appropriate policy was enforced by the former Colonial Government. A great deal of resentment, however, was whipped up by those who argued that this policy was part of a deliberate campaign to endanger the livelihood of cocoa growers and to impoverish the country and the fact that such rumours were mendacious and completely without foundation was of little account. There is little doubt that much of the unrest which preceded and probably accelerated independence was based upon disaffection and fears among cocoa producers, and this disaffection was further stimulated by the extremely fluctuating prices which farmers received for their crops (Fig. 18).

If we now move to 1966 we note that in this year the former President Nkrumah was overthrown by a military coup and it was surely no accident that the period leading up to that year saw steadily falling cocoa prices with consequent distress and apprehension among cocoa growers, all of which helped to give rise to political unrest. This is more easily understandable if we recall that in the early 1960s a Ghanaian farmer had to produce 3 tonnes of cocoa to purchase a new farm tractor manufactured in the U.S.A. or in Britain, but by the early 1970s 10 tonnes were necessary to earn the purchase price for a similar piece of equipment (*The Times*, 23rd May 1975). The whole financial status of the country had obviously deteriorated and there is little doubt that this deterioration helped to bring about dramatic political changes and another change of government.

Admittedly we have gone far from our starting point but this example does sharply remind us that when producers venture into the hurly-burly of cash cropping and all that this entails they may well find that they have left behind them any life of comparatively placid self-sufficiency which they may previously have enjoyed. The path of economic development is tortuous and rocky and no one who treads along it can tell whither it will lead.

PLANTATIONS

We should now return to our main theme, the development of cash cropping, and we may recollect Stage 5 in Winter's scheme above—the Agricultural Wage Economy which includes the plantation system. Much has been written about plantations and it is therefore perhaps surprising that even now there is by no means complete agreement about the exact nature of a plantation. Some writers, for instance, argue that plantations are concerned with tree crops but others would recognise the existence of tea, sugar, cotton and tobacco plantations. Some writers have suggested that a main characteristic of the plantation is the *in situ* industrial processing of its product before sale; thus the enterprise will sell sugar, not sugar cane, and tea, not raw tea leaves. But this definition will not cover all cases. For instance, cotton plantations do sell raw cotton (the ginning is not normally classed as an industrial process any more, say, than the threshing of wheat), while rubber plantations frequently sell latex though they do sometimes sell crêpe. Some argue that a plantation must be financed from foreign sources and that a fully home-based enterprise is not a true plantation but the economic and social effects of domestically based plantations are likely to match those of undertakings financed from overseas; the distinction seems arbitrary. Gourou (1968, 147) is quite clear that the plantation is an alien growth which produces a clash of cultures; "there is no need to talk of plantations when the clash of cultures does not occur." It can be argued that the clash of cultures appears, for example, in the regimented vegetational landscape which contrasts so markedly with the adjacent countryside; in the highly organised infrastructure (buildings, roads, perhaps light railways) which has no counterpart in the surrounding areas; and in the organisation and the techniques of the whole operation which are frequently dependent at least in part upon alien labour. If we accept this line of thought we are close to the thesis that the plantation is a cultural rather than a physical phenomenon.

The original reasons for the establishing of plantations are clear enough. When the first Europeans visited the tropics and sub-tropics they quickly realised that these regions could produce valuable commodities which would command high prices in their temperate homelands* but the difficulty was that there was no indigenous economic system capable of supplying these goods—at any rate on a systematic and reliable basis. The only answer to this problem seemed to be the organising of production by the Europeans themselves, and it was this organising which gave rise to the plantation. Land was secured and labour supplies were secured to meet the needs of the new form of production. Sometimes, as in the case of West Africa, labour was supplied by the indigenous people; sometimes, as in Malaya and Ceylon, it was imported from elsewhere; and sometimes, as in the Americas, the labour was organised on a basis of slavery. The necessary capital was provided by companies based overseas.

In view of the very unfortunate associations connected with this form of enterprise from the early days it is in a sense surprising that the plantation system has endured so long. The fact that it has done so strongly suggests that this productive system, whatever may be its other failings, has met a real economic need. It might be helpful at this point to mention some of the more important features which do help to distinguish plantations.

* Sugar, which was increasingly coming into demand, is an example of such a commodity and very high prices could be secured for it in the early days. The present writer (1974b, 13) mentions elsewhere the case of the Miles family whose fortunes were very largely founded upon sugar.

In the first place, the plantation is an extensive landholding. It is frequently owned by a foreign company which means that the necessary capital is of foreign origin. Skilled workers, machinery and equipment are also obtained overseas, the main contribution of the host country, apart from the land itself, usually being the supply of unskilled labour, though this has not invariably been the case as we have already noted. The pattern of foreign management and indigenous labour has frequently led to the eruption of ill-feeling towards the plantation from the inhabitants of the territories concerned, the more so since money profits have traditionally been remitted overseas.

On the other hand it should be borne in mind that the host country may expect to receive substantial benefits from a successful plantation. The government will certainly gain heavily from taxes and royalties while workers who are nationals receive wages. Plantations directly encourage the development of better transport services while they frequently provide health and education facilities for their workers and their families; in these days, moreover, nationals can secure advancement to the most skilled managerial and technical posts.

In many ways a well-run plantation is a most efficient economic unit; it must be if it is successfully to meet competition from other regions. Locations in near-coastal areas are favoured in order to minimise transport costs while technical skills among nationals are encouraged largely through the maintenance of plant and the processing of the plantation product on the spot before despatch prior to export. The maximum use can be made of modern research so that a better-quality product can be developed perhaps, for instance, as a result of botanical experiment, while the marketing of the product must be efficiently undertaken. All these activities mean that a wide variety of managerial, technical and research skills are introduced into the host country and this cannot but prove beneficial.

Despite all these advantages, however, there are many who argue that local small-scale producers need not today find themselves at any disadvantage as against the plantation, and they will point, for instance, to the success of West Africans in establishing the production of cocoa and to the many smallholders who together are responsible for two-thirds of the rubber output of Malaysia. Other success stories of a similar nature come from Burma and Thailand (rice production), from Colombia, the Ivory Coast and Togo (coffee), Bangladesh (jute) and Uganda (cotton).

It is to be expected that the plantation is often placed at a disadvantage because of the natural association in the minds of local people with the "bad old days" of slavery and colonial rule. Even today this feeling is often fostered, though not deliberately, by the employment of expatriates in senior positions as against the recruitment of local labour for unskilled and semi-skilled posts, and also by the remission overseas of profits, though these features are by no means as dominant today as they once were. When emotions, as opposed to rational thought, however, are involved features such as these can still be of over-riding importance in shaping attitudes in people's minds.

It is not surprising in view of such unfavourable circumstances that the plantation today not infrequently rests on an increasingly insecure foundation. This was demonstrated, for example, during the unsettled years in Indonesia when for some time after independence was achieved in 1949 the country went through a very difficult period characterised by a lack of physical security, unsettled labour conditions and uncertainty regarding land titles. These confused conditions produced sharp losses of productive capacity through malicious destruction of assets (both plant and plants!), squatting, expropriation and consequent inadequate outside maintenance which in its turn resulted in further losses through natural causes such as the spread of blight and disease. Financial problems became acute for plantation owners, not least because the falling output was no longer sufficient to keep the expensive, highly-capitalised processing plants operating to capacity. If such a situation continues for long it no longer is worth while attempting to service and maintain sophisticated processing machinery, much less to expand production, and the viability of the plantation as an economic unit must sooner or later be lost.

Another relevant point is that the plantation necessarily fosters an ill-balanced form of agriculture since it is designed to concentrate upon the production of a single crop or upon a very small range of crops. This makes for a lack of output elasticity, and particularly in times of falling prices or during periods of economic transition this can be a serious disadvantage. Humphreys (1946, 112–13) believes that economic development in Brazil, for example, has greatly suffered because for four centuries the economy was dominated by a series of plantation-based monocultures—sugar cane, cocoa, tobacco, cotton, rubber and coffee. Monocultures are also notoriously susceptible to diseases and severe losses have been incurred for this reason especially in the production of cotton, cocoa, rubber and bananas. On the other hand it is argued, and probably correctly, that the small-scale producer who diversifies his interests over a wider range of food and cash crops is far better placed to meet varying market demands. He can fairly easily expand or contract the output of any of his crops and can vary his labour input according to changes in market demand.

AGRICULTURE—FORMS OF FARMING ACTIVITIES

Bauer and Yamey (1965, 94–5) point out that it is very difficult to compare the relative efficiencies of the plantation and the smallholder because different criteria apply in each case. For instance, trees are usually planted far more closely on smallholdings than on plantations and it is often argued that this is a result of lack of knowledge and consequent inefficiency on the part of the peasant proprietor for there is no doubt that under the plantation system yields per tree are considerably higher. But other criteria have to be taken into account. The relatively scarce production factor for the smallholder is the land. Apart from the land itself the cultivator has only small capital assets while wage costs for himself and his family are minimal. His main concern is therefore to maximise output per unit area, and this he succeeds in doing. The plantation on the other hand has more land at its disposal and the relatively scarce production factors are more likely to be capital and labour. It is therefore sensible under these conditions to maximise returns per tree and per worker. The point is that plantations and smallholdings employ different production techniques and comparisons are therefore very difficult to make.

Another problem which plantations may have to face is that presented today by organised labour. The smallholder has no problem in this respect, for the labour is supplied by himself and his family, but a different situation arises in the case of a capitalistic enterprise which employs wage earners, especially when trade union activity must be reckoned with as a powerful factor pressing wages upwards. Government interest also may have to be taken into account if, for instance, legislation is approved requiring employers to pay wages which do not fall below a minimum level and to provide various amenities for their staff. All these factors help to tilt the balance against the plantation and in favour of the smallholder.

It seems fairly clear that the plantation–smallholder issue is by no means clear-cut, a conclusion which we might logically expect when both plantations and smallholders continue successfully in operation. In both cases so much depends upon environmental circumstances, both local and global, that generalisation is very difficult if not impossible. It might help, therefore, if we briefly examine two contrasting examples of plantation enterprises, one a success the other a failure, in order to show how much depends upon circumstances which are not always obvious to the outside observer.

A Malaysian Example

Our first example comes from Malaysia. Before the plantation era the Malay Peninsula was very sparsely populated so that most of the land remained unused and empty, but in 1876 came Wickham's successful attempt to carry rubber seeds out of Brazil, the growing of rubber trees from these seeds at Kew, and the further successful transporting of the young trees to Singapore where the lesser-known but vitally important experiments of H. N. Ridley showed that sustained and continuous tapping of mature trees could safely be undertaken if carefully controlled. Until that time it seemed that tapping might well seriously injure or even kill the tree and that, if true, would have meant that rubber collecting would necessarily remain a forest gathering occupation. Ridley's experiments, however, showed that rubber could be produced as a "cultivated" crop and the first commercial plantings in Malaya took place in 1895. Further development was rapid.

It is not proposed to give here a descriptive account of the development of the rubber-growing industry in Malaya for excellent accounts are available elsewhere; that given by Fryer (1965, 178–85) can be highly recommended. It is our purpose to enquire why the activity proved so successful, and we can pick out four main reasons applicable to this particular case.

1. The location of the Malayan Peninsula proved to be in many respects extremely favourable. The region is located within the equatorial climatic zone and this type of climate (Figs. 2, 4 and 5), constantly hot and humid, is essential for the successful growth of the rubber tree, while the long, narrow, peninsula-like form of Malaya meant that communications both internal and external never presented the kind of problem met with, for instance, in Amazonia. No part of the peninsula is remote from the open sea. The great port of Singapore quickly developed as a natural focus of trade while Penang also became an important route and shipping centre. Finally, Malaya was near enough to both India and China to make the import of foreign labour quite practicable.

2. The time was ripe for the opening up of large-scale rubber production. Until the 1890s, the value of this commodity was modest, reflecting its limited range of uses, but towards the end of last century the new motor-car industry was beginning to make very substantial demands upon available rubber supplies. This was an essential feature of the enterprise for without large-scale demand there would have been no possible point in large-scale production.

3. The political links with Britain also favoured the new industry as these favoured the movement of capital and skilled labour into Malaya as well as expediting the recruitment of indentured Tamil workers from southern India to provide the tapping force necessary. At the other end of the production line Malayan rubber was assured of preferential entry into the expanding British market.

4. There can be no overemphasising the contribution made by the research and the careful techniques which were employed in production. Some of these techniques we now take for granted but that is very largely because their value was demonstrated on the rubber plantations of Malaya. We have already referred to Ridley's painstaking work which made the whole venture possible but we might also mention the successful use of chemical fertilisers and the suppression of the mosquito even before the advent of modern insecticides. The former was a tremendously important step as tropical soils do not respond to chemical fertilisers in the same manner as temperate soils; sometimes, indeed, applications of these can be positively harmful, for instance when phosphatic fertilisers give rise in the soil to harmful aluminium phosphates. The mosquito problem was dealt with by careful water control, for instance, by channelling the streams in concrete flumes shaded by bushes and by taking extreme care never to allow any standing water to act as a breeding ground.

Another vital point was that the formerly widespread practice of clean weeding was completely abandoned and a cover of natural vegetation was retained over the soil. Neither ploughing nor wholesale deforestation were practised, the young rubber trees being planted in holes after the removal of only the largest trees. After careful treatment the natural vegetation of the forest became the undergrowth of the plantation and both topsoil and humus were conserved. In effect, the natural forest was gradually replaced by a largely artificial one.

There is no doubt that this is a success story. There is equally no doubt, however, that the day of the imperially based plantation such as we have been examining is over and its very success in past years has left problems behind it. Natural rubber accounts by value for about 30 per cent of Malaysia's exports and for 17 per cent of the Gross Domestic Product and this means that the country is very vulnerable to the markedly fluctuating prices which form an abiding feature of the world rubber market. One line of approach to this problem is for Malaysia, Thailand and Indonesia, which together produce about 75 per cent of the world's total natural rubber, to agree jointly on a price stabilisation scheme with the establishment of a "buffer" stock of about 10 per cent of the annual world production to help in evening out the fluctuations of supply. Such a scheme is likely to be set up though it may not prove a final answer to the problem any more than the Stevenson Restriction Scheme (1922–28) was; one main result of this earlier scheme was to encourage production in territories not parties to the agreement, for instance the Dutch East Indies (which later joined the scheme), Borneo, Thailand, Indochina, Brazil and Liberia.

Malaysia is also trying to devise schemes to cope with the domestic problems bequeathed to her from her imperialistic past. Even apart from the vital fact that rubber is the mainstay of the economy there exists the situation which is very difficult for an emergent territory to accept that only 0·3 per cent of rubber production is in the hands of Malays while resident Chinese own 13 per cent and foreigners 78 per cent. The remainder is held by the corporate sector. This unbalanced situation leads inevitably to considerable friction between the Government and the foreign business community and schemes are being considered to Malayanise the rubber estates and eventually to re-distribute them to smallholders. This is a formidable undertaking and there is always the fear that the implementation of such a scheme must result in a serious drop in efficiency and output, thereby weakening the domestic economy of the country as well as her status in world trade. There are no easy answers to these problems.

A Brazilian Example

We can deal with our second example fairly briefly. In 1927 the Ford Motor Company purchased about 1 million ha of land in Brazil along the right bank of the Rio Tapajos about 220 km upstream from its confluence with the Amazon, and about 3400 ha of this were planted with rubber trees; the name "Fordlandia" was given to the plantation and to the settlement which was established to service it. This venture was part of the attempt made by American interests to secure rubber supplies independently of the Stevenson Restriction Scheme. The enterprise, however, ran into difficulties almost immediately. In the first place the natural vegetation cover was mistakenly removed and soil erosion quickly became critical, while the ensuing silting which took place along poorly-drained valley floors meant that expensive drainage measures had to be undertaken. At the same time expensive terracing and the belated planting of cover crops became essential. The rubber seedlings which ironically enough had been imported from Malaya took time to become acclimatised and were attacked by a kind of blight. Finally, the Tapajos itself, despite its enormous overall length and volume, proved to be poorly suited to transport, particularly during the dry season the length of which increases noticeably upstream especially in the upper course of the river in Matto Grosso. This means that the river level falls very considerably at this time of the year.

In 1934, therefore, the Company exchanged a part of its Fordlandia concession for land at a new site, Belterra, only 50 km from the Amazon itself, where the terrain was more gently sloping, where the more friable soil made cultivation easier and where the river was more navigable. Even so the venture did not prove successful and the scheme was abandoned in 1946. The whole project had cost several millions of pounds but it was sold to the Brazilian Government for £80,000.

Interestingly enough, the difficulty which finally broke the enterprise was that of labour. The Brazilian workers did not take kindly to the routine work upon which the success of the project depended while the sparse population of the region meant in any case that the plantation could never attract enough workers. In a desperate attempt to attract labour the Company offered high wages but this policy foundered since the employees, used to a frugal way of life, saw no point in working a full week when they could earn enough to supply their modest needs by working for much less than that. Absenteeism therefore became rife and the labour problem remained unsolved. One remarkable result of this disaster is that Brazil, the original home of *Hevea braziliensis*, cannot today meet its own requirements of natural rubber.

A Case Study: Caribbean Banana Production

The whole question of the status of the plantation is clearly a pressing one today and it might be helpful to examine more carefully one example to show how this type of enterprise is adapting itself to changing conditions. The example chosen is that of banana production in the Caribbean region.

It is estimated that a greater tonnage of bananas is produced annually in the world as a whole (probably more than 20 m tonnes) than of any other fruit except grapes and possibly mangoes. They form an essential item of food to dwellers in the humid tropics and there are literally hundreds of different kinds. Some are large, some small; some are yellow, some red; some are sweet and edible when ripe while others are starchy and must be cooked; still others yield valuable fibres such as Manila hemp. The fruit is never out of season, which further adds to its value. Yet despite this wide range only a few types are grown for export in any quantity, easily the most important being the well-known Gros Michel—fat Mike—though the proportional importance of this type has declined in recent years. In fact only about 16 per cent of the total world crop is grown for export and production for this purpose is restricted to a few areas because the banana plant is very exacting in its requirements. It needs more than 2000 mm (80 in.) of rain a year with no dry season while temperatures should range between 24°–27° C 75°–80° F) throughout the year. Plenty of sunshine, however, is essential. These are, of course, optimum conditions and under certain circumstances the fruit can be successfully grown outside the limits which they set; it responds well, for instance, to irrigation in areas of less rainfall if the necessary copious supplies of water are available and if temperatures are consistently high enough. High winds can prove disastrous, not so much during the growing period as when the heavy fruit bunch is maturing; by the time that the fruit is ready for harvest and the bunches are at their heaviest the supporting stem has become so weak that even a 30 km per hour wind can flatten the whole plant with disastrous consequences for the fruit. Finally, the plant needs fertile, friable, well-drained soils with a clay content of less than 40 per cent. It is perhaps surprising that although the water requirements of the banana are so high it needs good drainage, and drainage ditches form a conspicuous feature of any banana plantation. There should also be a fairly high lime content in the soil as acid soils accelerate the spread of disease, especially the virulent Panama disease which attacks the roots and causes the leaves to wilt, while the main stem rots near the ground. An infested plant bears no fruit. Almost as devastating is Sigatoka, a leaf spot disease which, however, can now be controlled by spraying though control until recently was very expensive.

The most suitable type of environment for the commercial production of bananas is the equatorial or sub-equatorial lowland, and the main exporting areas are Central America (about 28 per cent of the total world supply), certain northern parts of South America (39 per cent, of which Ecuador alone accounts for almost 28 per cent), the West Indies (8 per cent) and the Canary Islands (3 per cent). In this study we are concerned with the Caribbean producing area shown on Fig. 19, the lowlands of which were the world's major producers until just after the Second World War although before the end of the third quarter of last century they remained largely untouched verdant sub-equatorial wildernesses with very few inhabitants.

The first signs of change came in the 1870s with the construction of a railway from Limon, on the Caribbean coast of Costa Rica, to

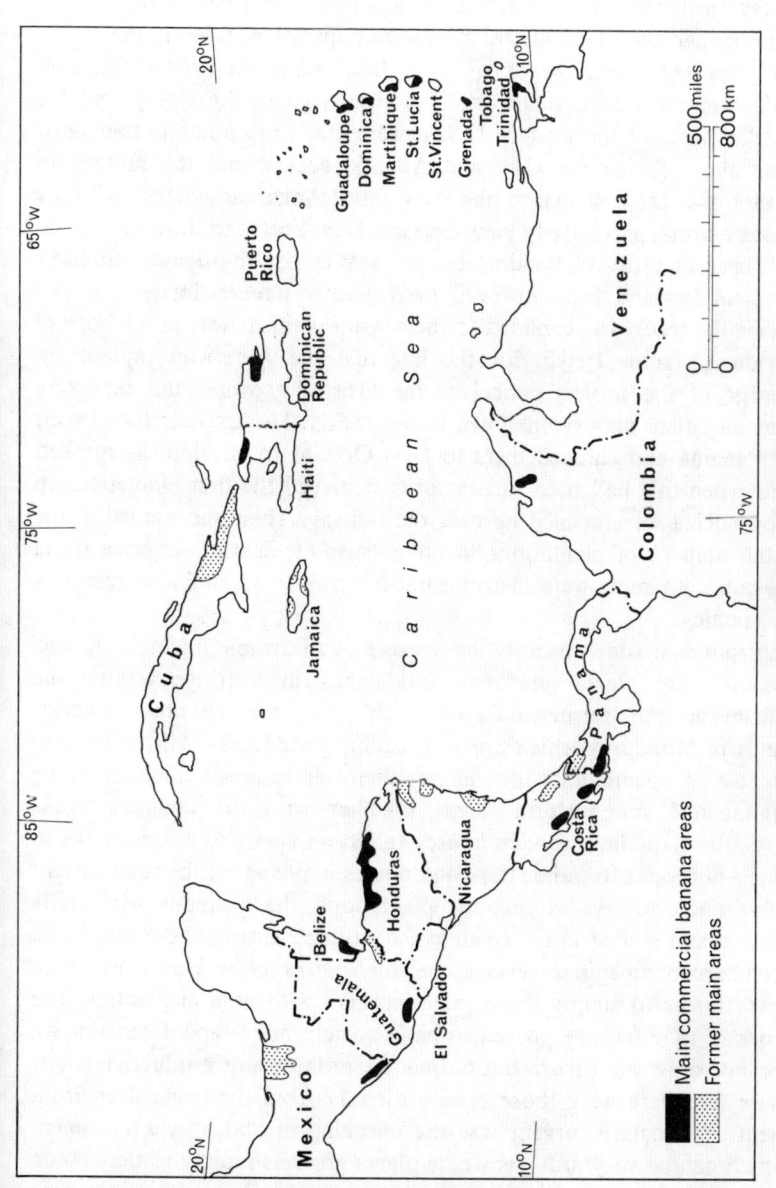

FIG. 19.—Banana production in the Caribbean.

San José on the spinal uplands of the interior. San José, the capital of Costa Rica, lies in one of the upland flat-floored basins which even today form the most densely populated parts of Central America; a century ago they held almost the entire population. Even in the 1870s the rich farm country of the San José basin exported coffee and hides *via* the Pacific coast and San Francisco to the U.S.A. but the market areas of the western U.S.A. were much more lightly populated than those facing the Gulf and Atlantic coasts, and the railway to Limon was the first step to link the Central American uplands with the eastern ports, particularly New Orleans, New York and Boston.

The difficulties of building the railway across the densely forested tropical lowland had, however, been greatly underestimated and for years the track was confined to the coastal plain. It was in the hope of producing some freight for the line that the American engineer in charge of the project conceived the idea of growing and exporting bananas, then almost unknown in the U.S.A. He first purchased fruit in Panama and shipped them to New Orleans to develop the market, and when this had been satisfactorily achieved the first plantations in Costa Rica were opened up near the railway. Their success led to the establishment of plantations in other parts of Central America while Boston merchants were instrumental in opening up similar enterprises in Jamaica.

From that time onwards the success of the whole undertaking was assured and plantations were established in a fringe around the Caribbean (the "former main areas" shown on Fig. 19, plus the coastlands of Honduras which are still leading producers). All the impedimenta of plantation life made their appearance including the regimented, monocultural areas together with the complex infrastructure including wooden houses raised on concrete pillars, tanks to catch rainwater for alimentary and domestic purposes, the roads, tramways and railways serving the plantations, the hospitals with staffs trained and skilled in the treatment of tropical diseases, the social and commercial amenities serving the alien *corps* of workers, and even dairy farms to supply these same workers with milk and butter. The work of maintenance proved to be extremely heavy and it remains so, for not only were there the normal hazards of any productive enterprise but there were those arising directly from the tropical environment. Particularly urgent was the checking of "bush" encroachment which can be very arduous where plants grow as rapidly as they do on sub-equatorial lowlands, together with the incessant clearing of drainage ditches and repairing of roads and tracks after fierce rainstorms. Such an enormous commercial undertaking inevitably required very heavy capital investment and it was natural that large-scale enterprise such as the famous United Fruit Company should quickly become dominant.

In the last resort the success or failure of the banana plantation depends upon extremely complicated logistics because the banana fruit is a very perishable product and it must be at the retail selling points very quickly after being harvested. Bunches are harvested while the fruit is still green because if it ripens on the tree it quickly becomes dry and insipid while the skin is also liable to split. They are then rapidly transported in padded trucks, by aerial cable or by tramway to a washing station after which they are despatched by rail for quick delivery to the loading port prior to export. Even these preliminaries require careful organisation. An incoming banana vessel will radio its approach and its cargo requirements from out at sea, and between the despatch of the message and the berthing of the ship a full cargo will have been harvested and moved to the quayside ready for immediate loading. Such is the military precision of the operation that within about 15 hours after the receipt of the incoming ship's message up to 80,000 bunches will have been harvested, cleaned, despatched from the plantations, loaded into the vessel's holds and begun their journey to northern markets.

Amid all this activity the destination of the ship must also be taken into account because this helps to determine the quality of the fruit harvested. Slightly more mature fruit, for instance, will be selected for the 5- or 6-day trip to New Orleans than for the longer voyage to New York or Boston while a thinner grade is normally harvested for a 14-day passage to England. On the journey great care is taken with regard to temperature control while sophisticated types of equipment are used in the U.S.A. itself to ensure the rapid despatch of the fruit from the ship to the inland distribution point near the final destination. At the terminals, for example, up to 8000 bunches per hour can be transferred by conveyor belt from the ship's hold to freight wagons or other carriers. Freight wagons are not only fully ventilated but are warmed in winter and cooled in summer to ensure that the fruit remains in good condition while in transit while on long train journeys into the interior even weather forecasts are carefully studied in case of environmental temperature changes along the route. It is not always fully realised what a complex organisation is necessary to ensure the success of the whole plantation enterprise.

Since the early 1940s there has been a very considerable migration of banana plantations from the Caribbean lowlands to new areas with the important exception of northern Honduras which remains a leading producing area. These changes came about for physical, economic and political reasons but the chief factor involved was the spread of disease among the plants, particularly the dreaded Panama disease which is virtually untreatable, and Sigatoka which could not be treated successfully before the Second World War and which has until fairly recently proved to be very expensive to keep in check. The combination of these two major diseases in the old-established areas virtually forced the United Fruit Company to shift its production from the Caribbean sector to lowlands facing the Pacific, while other interests developed on Hispaniola and the Lesser Antilles. Also encouraging this move were deteriorating soil conditions (the banana makes very heavy demands upon soil) while production in Jamaica was slashed by a series of devastating hurricanes.

There were economic advantages to be secured from the Pacific location. The existence of a dry season, for example, though necessitating expensive irrigation means that the spread of Panama disease is far less rapid than in the old areas, while Sigatoka too is more easily controlled. At the same time the forest growth is not as dense or as rapid as it is in the Caribbean and it is therefore less difficult, and therefore less expensive, to clear and to keep in check. It is interesting to note that Ecuador, formerly a producer of very limited significance, has leapt ahead in recent years until it is now the world's most important producer. It is also interesting to note that in Ecuador most of the crop is produced by comparatively small-scale growers and not by the large corporations which so dominate production farther north.

It is clearly not an easy matter to draw up a balance sheet in any attempt to demonstrate in a definitive manner that such forms of economic enterprises as banana plantations have or have not been on the whole of benefit to the host countries. In any such attempt, however, the following points would have to be borne in mind.

1. Before the banana-producing enterprises were established on the hot, humid and unhealthy Caribbean lowlands these areas were almost entirely uninhabited because of the hostile environment. These areas are now opened up with a developed infrastructure and numerous settlements, and because of the experiment and research which has been essential we have learned a great deal about the ways in which man can come to terms with a humid tropical environment.

2. A significant proportion of the expenditure and equipment involved remains within the host territories in the form of wages, taxes, health and educational services and improved means of transport, while in some instances "spin-off" benefits are made available to the local inhabitants. For example the United Fruit Company

provides scholarships for children of its employees to study in Canada and the United States and it also subsidises the Pan-American School of Agriculture in Honduras, graduates of which are not allowed to enter the Company's service.

3. At the same time there is a destructive element in many plantations as they take so much out of the soil; it is not for nothing that the banana has been called a "soil killer." The United Fruit Company has made strenuous efforts to encourage the growing of food crops by peasant farmers on former plantation land but has met with small success. It takes many years for a banana growing area to recover its lost fertility after a plantation is abandoned.

4. Because of soil exhaustion and because of the depradations of Panama and Sigatoka diseases plantations must be abandoned after anything between 5 and 15 years. Banana companies must therefore possess very large reserves of land for future production. The ensuing rapid shifts from one area to another, even from one country to another in some cases, does not make for stable economic and social development; it is disruptive to local societies and can leave them worse off than they originally were.

5. For reasons which we have already noted commercial banana production in Central America almost entirely rests in the hands of expatriate companies whose first concern is the profitability of the enterprise. Locally, banana prices have normally remained low and independent producers could not therefore remain in the business unless they were subsidised by the companies. This has been done mainly through the purchase from the independent producers by the companies of fruit at prices higher than the prevailing local prices though these prices have customarily been low. In recent years, however, the United Fruit Company has adopted a much more enlightened policy towards local growers who are now guaranteed the cost of producing their crop. They also receive a share in the company profits.

It is not surprising under these uncertain circumstances that the future of plantations remains very much in doubt. It is easy to understand, for example, the image of the United Fruit Company in Costa Rica as a *yanqui* corporation which has exploited local resources and local labour and which is an entirely alien growth with no basis in and few contacts with local society and culture. These features make the Company a vulnerable target for politicians and reformers and threats of political interference must be viewed seriously by any company concerned and taken into account in any planning for the future. On the whole it seems safe to suggest that we may be seeing the end of the plantation system, certainly in its present form; recent experience in Ecuador to which reference has been made suggests that national interests may be able in the future to establish production units able to satisfy the requirements of overseas customers, and if this does happen the basic reason for the existence of the plantation will have disappeared.

LIVESTOCK FARMING

In general it is true to say that the humid tropics are not favourable to stock-rearing and the indigenous peoples frequently have little use for stock as part of their economy. It is often the case that it is comparatively recent immigrants who are pastoralists; within this category we might include the Fulani of West Africa and various Hamitic or semi-Hamitic groups such as the Masai in East Africa. It is comparatively common for cattle to be kept as a badge of prestige and power; the important item then is the number, not the quality, of the beasts. Gourou (1968, 75), for instance, refers to the instance of Madagascar, an island covering 596,000 km^2 (228,000 sq. miles) with a population of about 8,000,000 (1972 figure) and about 6,000,000 head of cattle. The importance of the cattle is traditional and social, not economic. Meat is not a normal item of diet, oxen have never been used for transport, and the only useful service which they render seems to be the treading down of flooded paddy fields before the rice seedlings are planted out. But a large herd confers great respect and prestige upon its owner and cattle certainly play an important part in social life. The Betsileo dialect has more than 120 terms for describing oxen according to their coats! Cattle frequently figure in the "bride price" which has to be paid to a father before the marriage of his daughter. Other examples of comparable situations are not difficult to find and that of India with its 200 million cattle, about 0·4 animals per head of the total populations, is well known.

There are various reasons why numbers of stock remain comparatively low in most parts of the humid tropics, one obvious one being that not all animals thrive in a constantly hot, moist climate. On the other hand a long dry season can lead to difficulties with regard to water supply for like human beings animals must drink at fairly frequent intervals in order to survive. Many useful animals, however, could do well even in the humid tropics given the right conditions and the fact that there are so few in most areas strongly suggests that other conditions are not in fact favourable. One widespread disability is of course the incidence of disease for not only are animals subject to infection as are human beings but it is impossible to accord to stock the same protective measures that can be enjoyed by people. To take an obvious example, it would not be practicable to protect sleeping stock with mosquito nets. On the other hand if immunisation measures can be developed animal populations frequently increase. Thanks to a successful campaign to immunise cattle against rinderpest in the years following the Second World War, for example, the number of cattle in the Gambia more than doubled in two or three years though a rate of increase of this nature could not for long be sustained. It is probably reasonable with regard to this present point to suggest that the conclusions which we came to earlier (p. 70) with regard to the relationship between disease and population in the humid tropics also hold good with regard to animal populations, and that with a greater degree of intensive land use and economic development we can expect the effects of tropical diseases markedly to diminish.

We should not overlook the fact that while the quality of indigenous stock (zebu cattle, local swine and poultry, for example) is frequently poor this deficiency can often be overcome by careful up-breeding involving the introduction of superior quality stock from temperate regions. The question of the reaction of the resulting mixed breeds to their environment thus becomes a matter of the greatest importance and is one which is receiving close attention at the present time.

Another difficulty of widespread occurrence is the poor quality of tropical pastures. Tropical grasses in general are low in nutritive value and are frequently unpalatable and the fact that they are widespread over wide areas does not make up for these deficiencies. One result is that stock fed on these pastures are slow in growth and in development and it also follows that the carrying capacity of tropical grasslands is low. A Malgash ox, for example, takes 6 or 7 years to reach maturity while it has been stated (Gourou, 1968, 64) that an acre of normal tropical pasture can support only one-tenth of the live weight of stock customary on European pastures (22 kg as opposed to 220 kg). This poor performance, however, can be greatly improved in many instances by controlled pasture management and grazing and it has been shown that in parts of Zaïre natural pastures which can carry only 1 cow to 7 ha can in this way be improved so that they can carry 1 cow to 3 ha. With even greater care involving the sowing of legumes it is even possible to graze 1 animal per 1 ha. It would of course be rash to assume that comparable results can be expected generally in the humid tropics.

Unfortunately there is a great temptation in communities which value stock for reasons of prestige to overgraze existing pastures and this inevitably leads to vegetational degradation and soil erosion. The first signs of erosion normally occur around drinking pools, along established tracks and in compounds (p. 49 above) where the feet of the animals quickly destroy any protective vegetation. The annual burning which is frequently necessary to maintain the pastures further damages vegetation and this can be just as powerful as the burnings of the cultivator in modifying natural vegetation. It has been argued that it is the annual burning of the Venezuelan *llanos* by pastoralists which maintains these grasslands and which ensures that they do not revert to the forests which almost certainly originally covered them.

It is important to note, however, that under existing circumstances such burning is essential in a tropical country with a well-marked dry season for it is the only practical way for the pastoralists to get rid of the old growth, while in some areas it actually improves the composition and quality of the pasture. Hodder (1973, 115) mentions that in parts of Africa a useful species of red cat grass is favoured by burning while other valueless species are suppressed.

It is very tempting to argue that cultivation could be made more efficient in the humid tropics with the help of animals—as draught animals for instance—but this in general is not a practical possibility for the simple reason that the shifting cultivator cannot normally produce sufficient food to keep stock strong and healthy even if other circumstances would permit. Haswell (1953) has examined this point with reference to a village in the Gambia in which an unsuccessful attempt had been made to introduce draught oxen into a hand-hoe economy, but the cultivators were quite certain that they could not produce enough food for themselves and also the oxen. And although the local oxen which were used in the experiment were derived from Ndama and Zebu strains with a predominance of Ndama blood which confers a relatively high degree of resistance to trypanosomiasis, there was a real risk that stock would succumb to this disease especially if they were worked hard. There is also a further point; the use of draught animals if efficiently employed can lead to disguised unemployment if additional land cannot be brought under cultivation, while the animals can become a tremendous burden if there is a lengthy dry season when farming activities cease (Fig. 15).

Comparable situations have been noted in various parts of the humid tropics, for instance in Indonesia (De Vries, 1954) and Nigeria (Grove, 1957) but on the other hand Hodder (1973, 118) argues that there are some parts of the monsoon lands with a fairly short but intense rainy season where the use of draught animals is essential for cultivation. The reason for this is not given but it may well be that the period available for the preparation of the land, including ploughing, is extremely short owing to the rapid onset of the rains, while the crops must be in the ground quickly as the rainy season itself is short. Under these circumstances the use of animal power may be essential.

It is an unfortunate fact that in the humid tropics there are only very limited means of storing produce for any length of time and this is particularly true of the special cold storage facilities which are essential for the storage of animal products such as meat, milk, butter and cheese. While this is a difficulty which could in theory be surmounted the financial burden involved would be very great indeed and it would be scarcely justified at the present time. The fact is that any large-scale introduction of stock into the humid tropics would present us with a very complicated set of problems to which there are no easy answers and it is very difficult to decide whether it would be advantageous even on purely economic grounds to encourage such a development. The Gambian example given above illustrates this point. And can we justify such a programme when we are faced with a world situation in which the ratio of agricultural land to present world population may be as little as 0·81 ha (2 acres) according to an estimate from the U. S. Department of Agriculture? If on this basis every person in the world received his fair share of food he would have only enough to provide himself with a diet of cereals little above starvation level. Such a diet is in fact the norm over extensive areas, including much of the humid tropics, and inhabitants of more favoured lands are only able to eat substantial amounts of high-protein foods such as meat, dairy and poultry products because the balance of economic strength is at present tilted in their favour. Is it, under these circumstances, *at present* right to encourage further extensive developments of stock rearing when the overall food situation is so precarious and when on average it requires about 7 kg of grain fodder to produce a single kilogram of meat, a ratio which also roughly holds in the production of the other high-protein foods?

It must in all fairness be borne in mind that global statistics such as those just quoted are frequently suspect as there is no definitive and objective method of arriving at them; we have come across this kind of difficulty in earlier chapters. The figures themselves and any deductions based upon them must therefore be viewed with some reserve and there is little doubt that some authorities will reject those given as they stand. Even so, the global situation is undoubtedly one of considerable potential danger and it must give rise, for instance, to serious doubts regarding the wisdom of present attempts to establish an extensive cattle rearing industry in the Matto Grosso State of Brazil. The rain forest is being extensively destroyed by felling and burning and pasture is sown to support the zebu-type cattle which can withstand the harsh conditions; these include tough grasses, a hostile climate and disease, including trypanosomiasis. After fattening the animals are slaughtered and the meat is despatched normally by air to distant city centres such as Rio de Janeiro and São Paulo. This form of land use is a monoculture for no attempt is made to diversify production and, equally, no measures are taken to conserve soil fertility or to guard against erosion.

Some writers have been keen to urge the extension of mixed farming within the humid tropics but the whole question is a very debatable one. The sheer burden of supporting large numbers of well-fed healthy animals would be enormous and there is no doubt that at present such a development is quite outside the bounds of possibility even if it is desirable. It is true that in temperate lands mixed farming gives excellent results but this is under conditions which are quite different from those obtaining in the humid tropics. Rotational crop farming and good management can result in very considerable increases in food output in low-latitude lands without the very heavy economic burdens which the introduction of mixed farming would entail. We are simply not as yet in any position to make a final judgment on these very controversial matters.

Chapter VI

Increasing Agricultural Productivity in the Humid Tropics

THE PROBLEM EXAMINED

THERE is much to be said for the view that the central problem facing the countries of the humid tropics today is what has been called the "demographic–economic problem" (Hodder, 1973, 119) which is compounded partly of population and population growth and partly of economic development. This problem was touched upon in Chapter III and it is well illustrated from the example of Java, an island which is mantled with a large proportion of unusually rich volcanic soils and which supports the densest population in the world (excluding such densely peopled fragments as Hong Kong and Barbados).*

Although Java comprises only 7 per cent of the total area of Indonesia it supports 78 m inhabitants, about 64 per cent of the total population of about 121 m, and it has a population density of almost 600 persons per km^2. This is a remarkably high figure for a predominantly rural island and it is not surprising that Java is the dominant part of the physically fragmented territory of Indonesia.

The Javanese *pe-tani* (peasant farmer) shows very considerable skill and initiative in his farming practices. He typically farms about 1 ha of land but he cultivates them with extreme intensity. With the help of terracing and irrigation (and thanks to the warm, humid climate) he raises two or sometimes even three crops a year. His farm comprises *sawahs* (irrigated rice fields), *tegalans* (dry farming plots) and a small area around his dwelling where he produces fruit and in which he frequently maintains a well-stocked fishpond. He is likely to sell rice, cassava and maize in local markets and with the cash thus earned he buys various commodities such as paraffin and other household goods.

---

* It should be made clear that some parts of Java do not have the advantage of these volcanic soils yet they are still densely populated.

There can be few areas of a size comparable to the island of Java which are so meticulously cultivated; it has been said that journeying through Java is like "wandering through an endless, carefully cultivated garden" (Higgins and Higgins, 1963, 20), with villages shaded by canopies of palms, pawpaws, mangoes and bananas rising out of the rice fields "like cool, green islands." Yet despite this skill and industry and the inherent fertility of the soil *per capita* incomes remain distressingly low, and the gross domestic product (*per capita*) is probably barely $70 *per annum*. One main result of this is widespread indebtedness and the development of a kind of *métayage* system. The smallholder rents out his *sawah*, the rent being paid to him in advance on the understanding that he cultivates the field and hands over a proportion of the rice crop after the harvest to the *rentier*. It may be noted that personal indebtedness is a normal and widespread feature in the humid tropics.

The demographic problem with which we are concerned has been carefully examined in Chapter III and we shall in this chapter examine the related problem of increasing agricultural productivity in the humid tropics, though the starting point of our investigation has already been referred to. Traditional societies such as those with which we are concerned for the most part operate a closed economic system, that of subsistence farming based on the shifting cultivation pattern, though in fairness we should remember that today there must be very few, if any, communities which are completely traditional in this sense. Some modifications of the old patterns have crept in as contacts have developed with the outside world. There is no place, however, within the strict confines of shifting cultivation for development or change; the system is self-perpetuating and completely static. We saw in Chapter V, for example, how attempts to improve production with the use of draught animals frequently come to nothing while at the same time the dead hand of tradition and custom often throttles energy and enterprise.

The immediate result of this situation is a low prevailing level of agricultural productivity, whether measured on a *per capita* or a per unit area basis, which is just another way of saying that crop yields from most parts of the humid tropics remain depressingly low. There are many reasons for this, but fundamentally it is a question of inefficient land use, and it does not stem simply from the presence of inherently inferior soils as is sometimes supposed—though the existence of these in many areas cannot be doubted. Thus Gourou (1968, 17), in referring to the superior yields of crops in many temperate, as opposed to tropical, areas, is careful to point out that farmers in temperate regions frequently tend their soils carefully, applying organic and inorganic forms of fertilising compounds, while the farmers of the humid tropics rarely do this. But tropical farmers suffer, and this is the core of the problem, from a crippling scarcity of capital though labour is for the most part abundant and cheap. The result is that labour-intensive forms of production are employed though *per capita* outputs remain low as in the case of the Javanese example given above. We can better appreciate the points involved if we follow for a while the path of the economist who recognises that certain *factors of production* are required for any form of productive enterprise, including agriculture. These are land, labour, capital and enterprise and in general the terms are self-explanatory. We might, however, note the following preliminary points.

Land includes not only the areal surfaces over which productive activities take place but also the soils and the underlying rocks. Thus, minerals are included within the terms of the definition but not the crops which have been introduced into the soil. There is a case, however, for including natural vegetation within the scope of the term. It is important to note that the supply of land is not fixed in amount as the early economists argued. Additional areas of land have been made available by various reclamation schemes (the case of the Red River delta was mentioned earlier (p. 70)), while the polders of the Netherlands form another well-known example. In a sense additional land is created whenever, say, the ground upon which a factory or office stands is duplicated by the building of two or more storeys and the same is true whenever previously unused land is brought under some form of production. On the other hand land is frequently destroyed by wrong usage. Cases in point arise from the incorrect use of irrigation which can produce saline deserts, as happened in parts of Pakistan, and from soil erosion which can render previously productive land valueless. Land which is comparatively useless for one purpose (*e.g.* for cultivation) may be valuable for another (*e.g.* for the building of factories or urban centres).

It is of very real importance that land is relatively scarce over much of the humid tropics, a fact which might cause some surprise to those accustomed to thinking of the apparent emptiness of so much of the region. Two things must, however, be borne in mind. The first is that under the shifting cultivation system a village community needs a very great deal of land on which to support itself. If, to take a possible example, a 10-year rotational cycle is followed, ten times the amount of land actually under cultivation in any particular year will be needed simply for crop production though to the uninitiated eye the nine-tenths will appear empty and unused. The second point, to which we return in another connection in Chapter VII, is that an apparent natural resource only becomes a real resource when it is actually available for use, and we have seen that with existing limitations shifting cultivators simply cannot cultivate extensive areas even if they physically exist. In such a case the land is there but it is not an available resource.

The relationship of this to our earlier point will be clear, that in a sense additional land is produced whenever previously unused land is brought under some form of productivity—perhaps a farming type of economy. The main point of this piece of reasoning is to demonstrate the falseness of Ricardo's outmoded idea that the supply of land is fixed by nature; the matter is not as simple as that. Neither, as we have seen, is land "indestructible" as Ricardo supposed for by wrong usage it can be degraded or even destroyed as an available resource.

Labour, to an economist, comprises the supply of bodily resources, both physical and mental, which are available for engaging in any form of productive enterprise. It should be noted that a productive enterprise in economic thought is not simply one which makes goods. Equally productive are the service industries and on this broader definition workers such as administrators, teachers, doctors and bookmakers all supply labour in their different spheres of activity. This is, in fact, a complex subject which cannot be fully discussed here but in general we may say that labour is productive from the point of view of the economist if it commands a price which is usually known as a wage, salary, stipend or fee. In the humid tropics this definition would have to be extended to cover the efforts of the subsistence farmer who is not normally paid for his work but who is obviously a producer.

It is important to observe that labour can be of many kinds. It can be unskilled (or non-specific) semi-skilled, or skilled (specific); it can be physical or mental in type; it can be inefficient or efficient, expensive or cheap. The last point is perhaps the least generally understood. Labour is expensive if the wage:output ratio is high and cheap if it is low. This is very important with regard to the true cost of labour which is not cheap simply because wages are low. If a very low rate of output goes with the low wage the labour concerned is in fact expensive; conversely, if output is high wages can be high yet the labour may be properly viewed as comparatively cheap.

Capital is a greatly misunderstood term. Properly used it refers mainly to the stock of producer goods available for use in economic forms of production.* Such goods can be very crude and cheap or they can be very sophisticated and expensive. Thus, for example, a

---

* The economist also views as capital any store of material held by a producer, for instance the raw materials which he will ultimately use, and also the store of finished products which he also holds before they are finally sold.

subsistence cultivator may use a simple digging stick which he has made himself from the branch of a tree and a stone (which is used to add weight to the implement). On the other hand an oil refinery may be installed at a cost of some millions of pounds near an oil-importing estuary or an oil-field. Wide as is the gap between these two items of equipment they have at least two important features in common.

1. They make the production of consumer goods more indirect but more efficient. In theory a cultivator could manage without his digging stick and produce his crops simply by using his bare hands but this method would be less efficient and it would result in a loss of production. It would not, however, be possible to refine mineral oil without some form of capital equipment though early refineries were simple affairs with crude but cheap plant; they were extremely inefficient when compared with today's sophisticated installations. As the refinery has become more complicated and expensive it has become much more efficient.

In general it is true to say that the more refined and sophisticated (and therefore more expensive) capital equipment is the more efficient it becomes. Thus, a spade is more efficient than a digging stick and a power-driven rotavator more efficient than either.

2. Some element of "saving" is necessary for their production. In the case of the digging stick the amount saved is small and is represented by the time taken in the construction of the implement. Two essential elements are involved. In the first place the time is set apart from other, perhaps more pleasurable, pursuits; in the second place the time is productively used for fashioning the piece of equipment. These two elements are of the greatest possible importance in economic thinking and they are known respectively as *saving* and *investment*. Saving is an essential preliminary to the formation of capital but savings as such are of no value for this purpose; they must also be productively used. This is what investment is—the productive use of savings for the manufacture of producer goods. This is perhaps clearer in the case of the oil refinery for in that case the initial saving will have been in money which will then be spent in the manufacture of the equipment and its installation; in other words, the money saved will be invested.

This leads us on to note that it is often convenient to consider saving, investment and capital in monetary terms as so much money saved, so much money invested, or the capital equipment of a firm as being worth so much money. Indeed many people probably think of capital as money but this is not strictly so; the true capital comprises the producer goods which money savings will purchase (or, of course, has purchased). It is even usual for the economist to include as capital the accumulated knowledge and skills of a community for these can be built up only slowly, sometimes over quite lengthy periods of time. Thus a skilled and experienced worker may justifiably regard his expertise as capital which, when invested as he carries on his trade or profession, will earn for him a financial return.

There are three fundamentally important points to remember about capital and capital formation which follow from what we have so far learned.

1. The fact that there must be initial saving before investment is possible means that we must curtail our present consumption for the sake of the future even if, as is very likely, there is no immediate reward. An economist might illustrate this in the form of a simple equation:

$$\text{Curtailment of present consumption} = \text{Savings} = \text{Capital available for investment}$$

2. Capital equipment does not last for ever; there comes a time when it is worn-out and must either be repaired or renewed. Furthermore, it will at some time become obsolescent and therefore of no further value. In a technological society very large amounts of savings must be invested *simply to replace worn-out or obsolescent capital*, without any overall increase in the amount of capital equipment. The term *net capital formation* is sometimes used to denote the amount of new capital assets in a community which are over and above capital depreciation over a given period. Thus, £1,000,000 saved during the course of a year will make possible the purchase of capital goods to the value of £1,000,000, but if during the same time machinery to the value of £500,000 wears out or becomes obsolescent (depreciates) the net capital formation is only £500,000.

3. Capitalistic production is incomparably more efficient than "direct" production which would condemn man to a near-brutish type of existence. It is important to note that when we distinguish between a "capitalistic" society on the one hand and a "non-capitalistic" (*i.e.* a communist) society on the other we are not implying that capital is used as a means of production in one society but not in the other. The capitalistic form of production is used in both. The distinction arises with regard to the *ownership* of the capital. In a capitalistic society most of the capital is under the private control of individual persons and companies while in a communistic society it is mainly under government control. Most societies today are

probably best described as *mixed* as some of their capital is privately and some publicly owned.

Enterprise is provided by the *entrepreneur* who, according to classical economic theory, owns and manages a productive unit. It is the entrepreneur who bears the uncertainties which go with almost any form of economic activity, and he may in addition provide some or all of the capital required to initiate and maintain the business. A successful enterprise can bring considerable rewards to the entrepreneur but an unsuccessful one can be financially disastrous for him.

Some economists today argue that enterprise is simply a specialised form of labour and should not be therefore viewed as a distinct factor of production in its own right. One important reason for this undoubtedly is the fact that today the small-scale business is in decline and over a wide range of the economy it has been replaced by the large-scale limited liability company or even by the nationalised industry. In the large company with its managers and various types of shareholders it is impossible to pick out any single person as the true entrepreneur, the one who is ultimately responsible for success or failure and who stands to lose disastrously if the enterprise fails. This is admittedly true, but it is doubtful whether it justifies completely abandoning the recognition of enterprise as a recognised factor of production; many genuine entrepreneurs still exist in small-scale businesses and their success or failure depend upon a great deal more than simple willingness to work.

AGRICULTURAL PRODUCTIVITY IN THE HUMID TROPICS

We are now in a position to return to our original line of thought which had to do with the reasons why peasant farming in the humid tropics, though labour-intensive, is normally inefficient and results in very low yields. It will now be clear that any form of economic production is achieved by combining the factors of production in varying ways for different purposes, but unfortunately the subsistence farmer's command over these factors is very imperfect. His land, for instance, is likely not only to be limited in amount but to be generally poor in quality; the labour available may be comparatively plentiful (though *see* pp. 98–9 above) but is likely to be of limited skill; the amount of capital at his disposal is almost certainly very small; while enterprise is often stifled by the system of cultivation (p. 96 above). Under these circumstances output is bound to run at a low level, and this situation will be modified only marginally if the input of just one of the factors of production is increased as happens, for instance, if peasant labour becomes more intensive. There is almost certainly no improvement in the quality of the labour, just an increase in the number of hours worked and this in itself is not sufficient to increase output to anything but a limited extent because of the continuing limited or inefficient use of the other factors of production. This is why as we saw earlier (p. 99) there is a sharp upper limit to the effective use of labour under the shifting cultivation system. The only way to increase farming efficiency is to increase the input of all the factors—and this in effect means changing the whole system of subsistence farming.

The fact that an increase in agricultural productivity in the humid tropics is desperately needed can hardly be denied though it is very difficult to estimate in quantitative form the gravity of the present situation. One estimate was given in the previous chapter but a more up-to-date one is now available following the meeting of the World Food Conference in Rome in November 1974. The estimated total world population was given at that Conference as 4000 m of whom a quarter (25 per cent) suffer from malnutrition (*cf.* Boyd Orr's estimate, p. 2) and of these three-quarters (roughly 20 per cent of the whole population, about 750 m in number) are near the brink of starvation. Yet it was suggested at the same conference that if existing agricultural resources were used efficiently a much greater world population than that of today could be fed—possibly one even fifteen times greater! While we may receive these estimates with considerable caution the message they strive to proclaim is clear enough: it should be possible so to increase agricultural productivity that all the inhabitants of the world for a long time to come have enough to eat.

It might be useful as we consider the implications of this to examine the various possible ways of increasing food output, using as our frame of reference the factors of production which were given earlier. If we adopt this line of approach we might base our study upon the following scheme.

Factor of production	*Points for consideration*
LAND	Land tenure
	Land use
	Land extension
LABOUR	Quantity of labour
	Quality of labour

CAPITAL	Use of fertilisers
	Crop improvement
	Water control
	Use of pesticides
	Storage facilities
	Mechanisation
ENTERPRISE	Initiative

The above scheme is not meant to be exhaustive or definitive. Some of the "points for consideration" have in fact been dealt with earlier but they are here referred to again from a slightly different standpoint and for the sake of completeness. It should of course be realised that topics of this kind are extremely complex and to cover them thoroughly would need a much longer treatment than we can here accord them, while the treatment of some of the points will be designedly brief as various technical, biological or botanical issues are frequently involved and these are outside the compass of an introductory geographical work. And, finally, it should be realised that it is sometimes difficult to assign a particular point to a particular factor of production. For instance, the use of fertilisers and pesticides is considered here under the heading "Capital" because capital resources are heavily involved in their development, manufacture and use, but questions of labour and enterprise are also involved. With these reservations, however, we might now proceed to examine the points set out above.

LAND

Land Tenure

This is a topic which has been discussed in Chapter IV where the important point was made that under conditions of communal land ownership such as are typical of societies which practise shifting cultivation individual initiative and enterprise are discouraged; this inevitably has the effect of holding down agricultural production. The problem is rendered more acute because the concept of communal ownership is by no means a simple one as it can allow several different kinds of right to exist over a single piece of land. The right to plant crops, for instance, may belong to one family but the right to pasture may be held by all the members of the extended family group; the right to any fruit from standing trees such as the oil palm may belong to one person while the right to hunt over the land may be vested in the community as a whole. There is no doubt that the existence of several concurrent rights over the same area makes it almost impossible for a farmer to initiate any scheme of land improvement. Any attempt, for example, to plant permanent crops or to enclose any plots would conflict with the rights of others. Sometimes a stranger is allowed to settle within the local group and he may be granted certain land rights on a temporary basis on the payment of gifts, often on an annual basis, to chiefs, priests and to families with hereditary rights of use. Such payments take the place of rent in monetarised societies.

In parts of South America small-scale cultivators (*peons*) have customarily been allowed to grow crops on the large estates owned by the *caballeros* (the land-owning classes) but since the *peons* did not own the land there was no incentive for them to initiate land improvements which might have led to increased production.

In many parts of the humid tropics land has traditionally been subject to religious laws. In such cases the farmer under customary tenure could do nothing without the blessing of the priest as well as the agreement of the family elders. Some African peoples have traditionally believed that the spirits of the dead retain a special interest in the land; a Nigerian chief expressed this belief when he said: "I conceive that land belongs to a vast family of which many are dead, a few are living, and countless numbers are still unborn" (quoted in Pedler, 1955, 29). These traditional beliefs have for some time been giving way to modern economic practices but their importance even today should not be discounted.

It is not surprising therefore that many writers have urged that the land tenure problem in the humid tropics is one of fundamental importance and that until the cultivators themselves have a more personal and direct interest in the land than they so often do today there is little hope for schemes of agricultural improvement. There are, however, four important points to bear in mind.

1. There is bound to be resistance to any change, a resistance which will arise partly from an innate conservatism and partly from fear. Bauer and Yamey (1965, 174–5) point out that the more urgent is the need for land tenure reform the more anxious and vocal is the resistance to it likely to be. For instance, if the land is becoming increasingly unable to support those living off it, perhaps through soil deterioration or because of a growing population, people will cling tenaciously to their customary rights and privileges as they will fear radical changes the value of which cannot be proved in advance. Times of poverty and doubt do not normally result in a willingness to experiment though they may result in civil strife if the point of

desperation is reached. Further, if some stand to gain from change it is more than likely that others stand to lose, at least in the short run, and those who are dispossessed may suffer much distress; the story of the English enclosures is packed with horrifying instances of the misery which those changes inflicted upon thousands of country folk, advantageous as the changes proved to be in the long run. If, of course, the re-distribution of land is at the expense of large landowners as was the case in Latin America there will be resistance from these powerful groups and in such a case a revolution may be necessary before the change can be effected.

2. The timing of any changes must be right. A premature change can effect little as the example of Mexico given above (p. 96) shows and it will in all probability simply give rise to considerable fear and distress with few compensating advantages. On the other hand prolonged delay may increase opposition to any change and thus ensure that when change does come it is more explosively violent than it would otherwise have been.

3. The land must not be too heavily fragmented after any scheme of redistribution for even under conditions of personal ownership this is likely to lead to uneconomic working with a consequent depression of crop yields. The farming of dispersed fragments of land is usually uneconomic because time is wasted in movement between the plots; duplication of capital equipment may be necessary; the use of large and more efficient equipment may not be possible; control of weeds and pests is more difficult; and it is not easy to experiment with new crops and farming methods.

4. Revision of land tenure will not of itself bring increases in agricultural productivity, which depends upon quite different things. All that it can do is to provide a suitable framework within which more skilful and enlightened farming practices can be carried on.

Land Use

Better forms of land use are widely needed in the humid tropics, a point which has received considerable attention in earlier pages of this book. Although we must bear in mind warnings such as those given in the last chapter, for instance regarding the too-enthusiastic introduction of cattle into the humid tropics, forms of crop rotation or sometimes even mixed farming are badly needed to take the place of all-too-prevalent practices which are frequently monocultural and which result in soil degradation and even destruction through heavy erosion. In Malaysia, for example, the continued cultivation of cassava and pineapples on cleared forest lands has almost ruined extensive areas for no attempt is made to fertilise the soil and none to protect it against erosion which becomes therefore progressively worse. The invasion by *lalang* grass with its rank growth and its consequent propensity to catch fire and to burn fiercely during the dry season prevents the re-colonisation of the clearings by normal types of secondary vegetation. Such areas are likely to remain little more than man-induced deserts for a very long time to come. A similar situation has developed in parts of the Rice Economy Belt of West Africa (Fig. 16) where large patches of non-productive land can frequently be seen, especially in Sierra Leone. These patches have been ruined by the too-frequent cultivation of upland rice; most of the impoverished soil has been removed by erosion and the clearings are now floored by the rock-like *carapace latéritique* with only the thinnest of residual soil coverings.

A carefully-chosen pattern of crop rotation can be of great value in maintaining soil fertility for different crops make different demands upon the soil. Mineral requirements, for instance, vary while some crops have longer roots than others and therefore feed at deeper levels. A period of fallow is often helpful. Robequain (1957, 97) mentions that on sloping land in Java a rotation consisting of upland rice, tuberous or rhizomous crops, groundnuts and vegetables is frequently used while White and Gleave (1971, 78) state that the Cabrais people of Togo maintain soil fertility by the introduction of leguminous crops into their rotation and by the careful use of manure and household waste. The Manding people of the Casamance River basin in Senegal practice a rotation consisting of Guinea corn mixed with groundnuts until the soils are almost exhausted, after which bulrush millet and *acha* (hungry rice) are sown in successive years. On land situated farther from their villages they plant millet, groundnuts, Guinea corn and *acha* in rotation. The Dikwa, who live on the Mandara Mountains along which runs part of the border between Nigeria and the United Republic of Cameroun, have developed a more advanced farming system (White, 1941). They have constructed dry-walled terraces on which crops are grown in a 3-year rotation; beans with some millet, millet with some beans, and Guinea corn. Cattle are penned at night throughout the growing season and their manure together with night soil and household waste is collected and spread on the fields. Useful trees, some of which provide fodder, play a part in this agricultural system which is very near to a true mixed farming economy. The Dikwa, thanks to these progressive methods, are able to cultivate their terraces for as long as 10 years in succession after which a year or two of fallow supervenes.

Tempany and Grist (1958, 105–6) emphasise the value of legumin-

INCREASING AGRICULTURAL PRODUCTIVITY 145

ous crops in any rotation owing to their power of fixing nitrogen in nodules on their roots. If livestock are also reared as part of the economy they can advantageously feed on the stubble while the roots, if later ploughed into the soil, increase the nitrogen content. Leguminous plants are also deep-rooted and thus help to bring useful minerals from the sub-soil to an horizon nearer the surface.

There is, of course, much that could be written regarding improved land use in the humid tropics but the examples just given will suffice to show the importance and the possibilities of this aspect of agriculture. It is, however, unfortunately the case that lack of labour sometimes limits the adoption of crop rotations while a community unused to the practice may well be reluctant to adopt it unless it results in obvious and rapid advantages. These may take the form of clear improvements in crop yields or in the production of a crop for which there is a strong market demand and for which a worth while cash return can therefore be secured.

Land Extension

This topic has been briefly touched upon above (pp. 135–6) and little more need be said here except to emphasise the main points concerned. On the one hand there is no doubt that in many tropical territories the proportion of land actually under cultivation is very small. Zaïre, for instance, has only about 1 per cent of its total surface area under cultivation (not including fallow) while the corresponding figures for Brazil and Burma are 2 and 22 per cent respectively. Higher figures are recorded only in the densely-populated parts of Monsoon Asia where, for example, the figure for India is 52 per cent. At first sight, then, it does seem that there is plenty of scope for expansion of the cultivated areas in the humid tropics.

On the other hand it must be recognised that the extension of cultivation in the humid tropics is not usually a simple matter and we have already discussed the reasons for this. We have, for example, argued that the transfer of populations from one area to another brings acute human problems in its train while we must always remember that the quality of the land supposedly available is just as important a consideration as its quantity. The mere fact that an area is uninhabited, or almost so, should not predispose us to imagine that it is necessarily just waiting for cultivators to come in and develop it. Neither does the voice of authority always give clear guidance in these matters. Hodder (1973, 135) refers to the estimates of Farmer (1957, 39) and the International Bank for Reconstruction (1962, 72) respectively with regard to the Dry Zone of Ceylon. Farmer, with the expertise of a skilled geographer, demonstrates that a careful examination of the problems posed by rainfall and water supply inevitably leads to the conclusion that the "empty lands" of the Dry Zone do not offer significant opportunities for settlement. The Bank, on the other hand, views the region as "the granary for the future" and as one which offers itself as a potential homeland for much of the rapidly increasing population of Ceylon.

It is difficult to discuss this whole topic in isolation as so many factors are involved, chief among which are the facts of the physical environment which are frequently far more complex than is sometimes imagined, while the relevant features of the human situation which themselves frequently constitute a complicated pattern of economic and social considerations are also very relevant. Not infrequently political considerations are also important with regard to attempts made to extend agricultural settlement and production, as we shall see later in connection with the notorious Groundnut Scheme of Tanganyika (as the territory then was). While it may well turn out to be the case that during an historical period of rapidly increasing population we shall see an extension of agriculture into areas which are today very lightly populated (*see*, for instance, the present opening up of the Matto Grosso of Brazil, pp. 131–2 above) we must bear in mind that such extensions often prove difficult and expensive and we might have to wait many years before we can arrive at any worth while assessment of their success or failure.

LABOUR

Labour is a crucial factor of production, for without it there can be no production of any kind. Even the plucking of wild fruits and berries involves a rudimentary kind of labour, while the hunting of wild game requires a specialised and skilled expertise for success. Even at this comparatively simple level therefore we find that not only is the quantity of labour a vital factor but also its quality, and this dual aspect of labour becomes more important with increasing economic diversity and complexity.

These questions have been discussed at some length in earlier chapters, particularly in Chapters III, IV and V, and it is not necessary to go over this ground again. It remains simply to emphasise that all schemes for human betterment will come to nothing in the absence of an adequate supply of labour endowed with the essential knowledge and skills. The economist is fully aware of this and he customarily draws a distinction between *non-specific* and *specific* labour. The former is provided by unskilled workers who can move from job to job without difficulty, while the latter is trained for and skilled in one particular kind of work. Both kinds of labour are vital to the well-being and development of a community but with an increasing measure of development the proportion of specific labour increases and as labour becomes more sophisticated workers demand more by way of reward, not only in terms of labour contracts and wages, but also in social and cultural amenities. While on general grounds this is to be expected and, indeed, welcomed there is no doubt that such expectations do make prospective expansionary plans more difficult of realisation, and to that extent they act as a drag upon progress. This may be particularly true in the case of many lands of the humid tropics for many highly skilled and cultured expatriate workers of the managerial and executive type may well be discouraged from taking up posts there because of the overall lack of cultural and social amenities.

It can scarcely be doubted that the rapid economic development which has characterised the industrial territories over the past two centuries would not be possible in today's climate of labour demands for high wages and amenities and in the face of legislation framed to benefit workers (p. 60–1 above). This is a powerful reason for expecting that future development in the humid tropics and indeed elsewhere may well be unable in the future to match the growth rates of developing countries in the past.

CAPITAL

In any farming community, and most communities within the humid tropics are of this kind, private investment by individual farmers is essential, even if the investment is limited to the provision of such improvements to the land as the provision of drainage channels, fencing or bunding. Such investment is frequently performed in the tropical lands by family or hired labour during the slack season and it is not therefore dependent upon large amounts of monetary capital. But for the kind of investment which will result in markedly increased prosperity, for instance the purchase of fertilisers, better quality seeds and various kinds of mechanical appliances for use on the land, money capital is essential. Unfortunately the peasant farmer by definition does not possess money capital on any significant scale while interest rates charged by money-lenders is normally exorbitantly high as can only be expected when small-scale farmers can offer no collateral and when the returns are so uncertain. In many parts of the humid tropics land is so expensive, even where it is possible to purchase it, that most villagers cannot ever hope to own the land they cultivate and they necessarily assume the status of tenant farmers. In much of southern India some form of share tenancy is usual, either *varam*, when the landlord receives four shares of the harvest to the tenant's one, or *kuttagai*, when the tenant guarantees the landlord an agreed amount of produce per unit area irrespective of the yield. Under the first system the landlord supplies manures and shares the costs of harvesting while other costs are borne by the tenant, but in the second case the tenant assumes all risks and expenses.

Systems like these which entail modest or low crop yields and high rents do nothing to help tenant farmers to improve their lot and much thought has been given to working out possible ways of escape from this closed circle. Many writers, for instance, have urged the necessity of setting up various schemes, particularly agricultural co-operatives, to provide comparatively cheap credit without which small-scale farmers seem doomed to an unrelieved sequence of inefficient production and poverty.

During the 1960s a great deal of attention was focused upon the possibilities of the "green revolution" which, it was hoped, would lead to a great increase in food output in low-level production agricultural countries such as those of the humid tropics. India in particular took up this idea and attempted to put the teachings of its proponents into practice. The use of fertilisers, pesticides and of high-yielding hybrids was involved, together with irrigation and careful water control; unfortunately, all these measures require large amounts of capital and few of the farmers of the L.D.C.s can afford these initial costs. Furthermore, the foreseeable prospects arising from the use of these varying techniques are to say the least uncertain and no guarantee of success can be given. Crop responses seem to vary greatly from place to place and from time to time; in India, for example, a series of improving harvests was sharply interrupted by that of 1974 which turned out to be a very poor year necessitating among other things the import of 5 m tonnes of wheat from the U.S.A. at a total cost of £320m.

Use of Fertilisers

It is frequently emphasised that the simple use of artificial fertilisers in the humid tropics can give very disappointing results unless the soil environment is favourable. In particular, soil structure, water and air supply are of the greatest importance in this respect. Many authorities believe that natural manures including humus give more beneficial results than artificial fertilisers, though this is challenged by others. The truth may well lie somewhere between the two extremes, and it is quite likely that both natural and artificial fertilisers have parts to play in soil improvement though the fairly recent sharp increases in the prices of artificial fertilisers must modify the situation enormously. These increases have come about partly through the near-monopolistic position of Morocco as an exporter of phosphates, a situation which that country used to quadruple the world price of this essential raw material during 1973. This in turn means that cost/benefit analyses must be sharply revised as the returns to farmers from applications of phosphatic fertilisers in many cases will no longer justify the financial outlay involved.

An essential point which the farmer must constantly bear in mind is that different types of fertilisers are needed for different situations. One crop, for instance, benefits from one type of fertiliser and another from a different one, always assuming that soil conditions are favourable. Requirements of different soils also vary considerably. The chief types of fertiliser are those which supply nitrogen, phosphates and lime, while other essential elements include sulphur, magnesium and iron but with a few exceptions these are of widespread occurrence in soils. All these minerals, however, are necessary for successful plant growth and Fig. 20 shows the considerable increases in cereal yields following the application of nitrogenous fertilisers that have been recorded under test conditions in India. Clearly the application of nitrogen to the soil was beneficial in these cases especially since the maximum recorded yields are even higher than those shown on the diagram. The most widely used nitrogenous fertiliser is sulphate of ammonia which also gives good results when applied to sugar cane, tea, cotton and rubber as well as to cereals.

Phosphorus deficiency occurs widely throughout the humid tropics (*see* p. 89 above). Experiments conducted in Nigeria have shown that soil deficiencies of phosphorus or sulphur severely limit the yields of groundnut crops but because the groundnut plant is able to meet its own nitrogen requirements no benefit is secured from the application of nitrogenous fertiliser. The best results accrue from the use of superphosphates. Lime, which supplies the calcium needed by all plants, is also deficient in tropical soils mainly because of its soluble character and it is frequently used in the humid tropics as a fertiliser rather than as a soil conditioner as is the case in temperate latitudes.

The whole question of the use of fertilisers is part of the whole

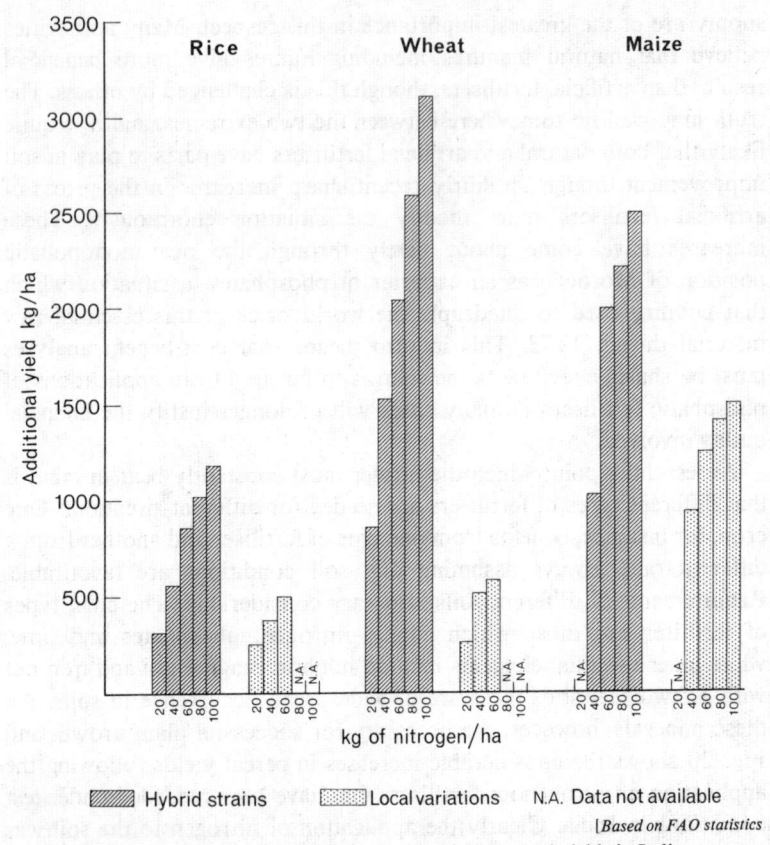

FIG. 20.—Effects of nitrogenous fertiliser upon cereal yields in India.
[Based on FAO statistics]

broader theme of correct soil use, a topic referred to in Chapter II. Not only is this a matter of maintaining soil structure and of applying the correct amounts of the type of artificial fertiliser needed for a particular purpose but there are also the important trace elements to consider. These are elements which need to be present in the soil only in minute quantities but which are nevertheless essential for successful plant growth. As in the case of the bulkier fertilisers different plants need different trace elements but the absence of these can produce a startling effect on soils, sometimes reducing them in quality so much that all they can support is poor heath and scrub. Examples of these elements include zinc, manganese and boron (*see also* below, p. 153).

Crop Improvement

The need for crop improvement in the humid tropics is a pressing one and we have previously referred to it in Chapters II and V. Such improvement depends primarily upon botanical research and experiment and it has involved the development of hybrid plant strains which can be higher yielding, disease resistant, pest resistant, or more tolerant of difficult physical conditions such as low rainfall. Attempts may also be made to improve selected physical properties of the crop, for instance to increase the oil content where appropriate or even to produce a plant of more economic size. In parts of West Africa, for example, oil palms can now be grown which have much shorter trunks than the original varieties and this means that the fruit can be harvested much more easily, quickly and therefore more economically. Not all these desirable qualities of course can be bred into any single plant strain; the research worker has to select the qualities which he desires and he must concentrate on them in his genetic experiments.

It is one thing to produce improved strains of economically useful plants, however; it is quite another thing to persuade farmers to use them. In part, of course, this is a matter of overcoming traditional practice and innate conservatism but it is often a matter of finance. Blanckenburg (1962) has shown with reference to eastern Nigeria that farmers frequently use a mixture of improved seed varieties with inferior strains for planting because the inferior seeds are much cheaper, and the farmers simply cannot afford to use the improved seeds entirely although yields from them are superior. In Sierra Leone it is reported that farmers normally use seed mixtures for planting their upland rice crops and these give low yields partly because the seed is inferior and partly because the different strains used in the mixtures ripen at different times so making efficient harvesting impossible.

The fact that it may be economically advantageous for farmers to use new and improved hybrids is strongly suggested in Fig. 20 which shows differences in yield under varying conditions between hybrids and local varieties of cereals. Not only are the newly-developed hybrids far superior in yield but they respond more readily to applications of fertilisers, and under favourable conditions they can increase harvest yields by as much as 200 per cent. This fact alone, however, will not be enough to win for them ready acceptance by farmers for they demand techniques of production which are not only new but expensive. We have seen above that peasant farmers often cannot afford the heavy initial costs involved, and a further disadvantage is that hybrids are frequently but unfortunately more susceptible to pests and disease; the risk of losses is therefore high—often too high to be acceptable. Liberal applications of fertiliser are also needed for the successful growing of hybrids and this further increases the capital costs involved, while essential irrigation costs and techniques are often beyond the reach of the average farmer. It is largely for these reasons that the "green revolution" has not been the success which was hoped for and, indeed, anticipated by many authorities and some research workers are beginning to consider possible alternatives such as the development of plant strains responsive to organic soil enrichment rather than to expensive artificial fertilisers.

Water Control

The reasons for the pressing and widespread need for careful water control in the humid tropics were examined in some detail in Chapter II where specific examples were given. We should, however, note that the practice of such control is not without risks and problems.

It is, for instance, in many cases tempting to imagine that beneficial results should be obtained from the draining of inland swamps, the presumption being that the deep soils which will be thus made available will be productive of crops of high yield. In some cases this may very well be true but one serious danger must be guarded against. The immediate aim of a drainage scheme must be to hasten the movement of water through the swampy area and this will probably be achieved by deepening and straightening the main channel and by running tributary drainage channels into the main artery. Such a procedure, especially if repeated at intervals along the course of a river and its tributaries during its passage through different swamps, can greatly aggravate the danger of flooding along the lower reaches of the river concerned.

Natural swamps act in large measure like gigantic sponges; during the rains they soak up vast quanitities of water much of which they slowly release during the dry season. If, however, a swamp is drained so that it does not absorb water during the rains it will have no water to release during the dry season and the problem of water supply downstream may well become more acute. In other words, increased rainy season flooding will be counterbalanced by acute dry season water shortage. Furthermore, the dried-out soils of the former swamp are quite likely to deteriorate beneath the hot sun of the dry season and the anticipated high crop yields may therefore never materialise—or if they do it is likely to be for a season or two only.

Use of Pesticides and Herbicides

The continuously high temperatures and high humidities which characterise the humid tropics favour the spread of a wide range of pests and diseases, while weed growth is apt to be prolific. Well-known examples include the stem-root disease in sugar cane which occurred in the West Indies towards the end of the last century; the swollen shoot and black pod diseases which have seriously affected cocoa production in West Africa since the Second World War; the sudden death disease which cut production of cloves in Zanzibar during the same period; and the rosette and black spot diseases which caused very considerable damage to the East African Groundnuts Scheme (*see* pp. 155–6 below), while we have previously mentioned the Panama and Sigatoka diseases which have played havoc with banana production in the Caribbean. At the same time insect pests are of wide occurrence and the effects of these can be most damaging. The pulse beetle, for example, passes through eight generations during the course of a single year and the offspring from an initial forty eggs can halve the weight of grain which suffers infestation in as little as 6 months, in addition to contaminating the remainder. The pressing need for weed control was stressed particularly in Chapter IV.

This is a topic of great complexity and we can do little more here than refer to it. Different problems require different solutions (if solutions there be!). There are circumstances, for instance, when the thorough cultivation of the soil can be beneficial as this can kill insect pests sometimes through crushing or by burying them and sometimes by bringing them to the surface to be destroyed by birds or the hot sun, while early planting has been shown to be helpful in combating rosette and maize rust diseases. Deficiency diseases can prove extremely troublesome and these can normally be remedied only by adding to the soil the particular mineral or trace element which is lacking though this treatment is only possible after careful soil analysis and experiment. Oil palms are susceptible to this type of disease, which includes orange frond due to insufficient magnesium and confluent orange spotting due to insufficient potassium. Lack of other trace elements such as zinc and molybdenum may also induce disease and so affect yields (Jarrett, 1958).

It has become normal in developed countries to protect growing crops with applications of pesticides and herbicides in the fields but in the humid tropics this practice is very limited and consequent crop losses due to insect pests, plant diseases and weed infestation may be as high as 70 per cent. Unfortunately the prices of pesticides and herbicides have at least tripled over the past 3 or 4 years and this must mean that the L.D.C.s will use them even less despite the better yields which they make possible.

Storage Facilities

The general lack of adequate storage facilities in the humid tropics has been referred to in Chapter V where some results were noted. There is also a close link between storage on the one hand and pests and diseases on the other, for poor storage inevitably brings substantial, if not complete, food losses through attacks by animal pests, insects and micro-organisms. It was pointed out at the Rotterdam Conference of the FAO, for example, that six rats can eat as much food as a single person and that in one year a single pair of rats can produce 70 offspring. A state of emergency was declared on Mindinao, one of the Philippine islands, in 1953 when rats destroyed about 70 per cent of the rice crop, while enormous losses have been reported from food warehouses as a result of rat infestation.

Mechanisation

We shall consider in this section the question of the substitution of machines for labour in the humid tropics, together with the related topic of the large-scale scheme. The two topics are not precisely similar but they frequently go together. This is perhaps especially the case in the L.D.C.s where small-scale farmers are normally poor and therefore cannot afford mechanical aids to production as their counterparts in more affluent countries often can.

The main advantages of mechanisation include savings in labour and time and the consequent possibility of more extensive cultivation than would otherwise be practicable. Thus, for instance, land can be tilled more effectively and more quickly by using mechanical help, while at the same time less labour is needed. It is frequently possible to bring under cultivation land which otherwise would not be used at all perhaps because the soil is too intractable for hand labour, the clearance too difficult or because insufficient labour is available. In fact the advantages of mechanisation seem so obvious that it is perhaps with some surprise that we come to realise that real problems are involved in its extended use.

If we lay aside at the moment the fact previously mentioned that agricultural machinery is expensive and therefore generally out of reach of most farmers in the humid tropics, we may recall that we saw in Chapter IV that labour-saving devices are not easily absorbed into the system of shifting cultivation. We must bear in mind that while mechanised production undoubtedly leads to great output per worker, output per unit area is lower, largely because hand labour is more meticulous than mechanised labour and land utilisation is therefore less intensive when much of the work is performed by machines. Fewer workers may be needed over any given area of cultivation if mechanised methods are used but total production is likely to be lower and this makes no sense where food outputs are very low to begin with and where there are no alternative jobs available for displaced agricultural workers. On the other hand there may be a clear overall gain in areas of low population densities and where there is only a short rainy season so that it is vital to prepare and plant the ground in a very short period of time.

It may be helpful to examine in some detail an actual example of a large-scale, mechanised agricultural scheme, as many of the relevant points will then be made clear. In fact the notorious East African Groundnut Scheme has proved to have been a classic example showing the pitfalls and difficulties which the large-scale scheme in the tropics is likely to run into. The scheme was initiated soon after the close of the Second World War under a Socialist government in Britain to increase the supply of vegetable oil available for consumption in western communities; "in other words, the motive was basically exploitative" (Hodder, 1973, 138).

Operations began in Tanganyika (as the territory then was; it is now Tanzania) in 1947. Three areas were selected for development under the scheme—one at Kongwa near Morogoro not far inland from Dar es Salaam, one much farther inland but still near the main railway at Tabora, and one in the far south-east of the territory near Nachingwea. The aim was to develop about $3\frac{1}{4}$ m acres of scrub by 1953 for the growing of groundnuts, mechanisation to be used whenever possible in order to speed up operations. In fact the original scheme proved a grandiose failure and the following are the main reasons (based partly on Hance (1964) and Stamp (1964)).

1. New machinery was not available in the immediate post-war years and much of the equipment used was therefore second-hand. This proved to be very unsatisfactory in operation as mechanical faults and breakdowns were frequent. Competent mechanics were not available in Tanganyika and this meant that a great deal of the machinery was out of action at any given time.

2. Bush clearance proved to be far more difficult than had been anticipated. The final clearing of most areas took four times as long as had been expected and it cost up to ten times the original estimates.

3. The tilling of the soil proved to be unexpectedly difficult. The abrasive soils blunted the ploughs until they were virtually useless

while the weight of the machinery compacted the soils until they became unsuitable for groundnuts which need a light, friable soil.

4. Harvesting turned out to be much less effective than expected as a large proportion of ripe nuts broke off the plants during harvesting and remained in the baked, indurated soil. Crop yields therefore suffered significantly.

5. Great losses were reported from diseases and pests. Rosette and black spot diseases, for instance, caused extensive damage to plants while wild pigs destroyed much of the crop.

6. Insufficient attention had been paid to the climatic situation and the first area (at Kongwa) to be developed proved unexpectedly dry. It is true that the situation was aggravated by a series of comparatively dry years but this is a feature to be expected in tropical regions.

7. The provision of social capital (housing, roads and other essentials) proved significantly more expensive than had been allowed for.

8. Local life and labour were completely disrupted by this massive invasion of capitalistic western-style enterprise. Wages increased sharply yet everyone suffered from excessive inflation. The traditional authority of the chiefs almost disappeared amid the hurly-burly of this example of capitalist promotion.

9. Management turned out to be inefficient and cumbersome. At one stage there were 1283 Europeans on the staff of the supervisory corporation but only 384 on the staff of the construction company which was actually doing the work on the ground. Most of the supervisors had previously had no experience of African conditions.

10. Transport facilities were totally inadequate while the railway from Dar es Salaam and the harbour there could not cope with the vast amount of equipment needed.

It is not difficult to see why the scheme failed. The failure was thrown into bizarre relief by the fact that during this same period in Northern Nigeria thousands of tons of groundnuts had to be stacked into "pyramids" because the existing Nigerian railways had not sufficient rolling stock and locomotive power to move them. A comparatively modest investment in the Nigerian railways could have resulted in the movement to the Western World of very substantial quantities of groundnuts without any great difficulty. It is difficult to avoid the conclusion that the main motive behind the East African enterprise was political expediency shown by a British government anxious to gain a point over its opponents.

It would be easy to point to other comparable schemes, few if any of which have lived up to initial expectations (some are described in Jarrett, 1974a, Chapter 17). The lesson of these examples is fairly clear; mechanisation cannot be effective unless it is used with discrimination and with full reference to local conditions and it will be economic only if it is applied to agriculture along with other appropriate techniques. Full account must be taken of the human as well as the physical background as the Kongwa fiasco so dramatically demonstrates.

ENTERPRISE

This is the last of the four factors of production and in some ways it is the most difficult to deal with. Indeed, as we have previously noted, some economists do not even recognise its existence in isolation from labour. There is a good case, however, for arguing that true enterprise as assumed by the entrepreneur possesses a quality which is different from that of labour. There is a readiness to assume responsibility for an economic undertaking, a responsibility which usually includes the provision of some or all of the necessary financial backing, and there is the knowledge that if the undertaking fails there may be severe financial losses. No worker has to assume responsibilities of this kind.

The great difficulty encountered in the humid tropics, as in all L.D.C.s, is simply that there is no tradition of economic enterprise and therefore there are in the first place none of the skills without which effective enterprise is impossible. Entrepreneurs can only come to the fore if the prevailing economic and social environment permits. There may be persons possessing the requisite ability and initiative among the inhabitants of the Amazonian selvas, and if these persons lived in a capitalistic society they would almost inevitably become powerful entrepreneurs directing the affairs of large-scale undertakings, but in the Amazon region there is no scope for these talents. And even if opportunities for entrepreneurship arose in the area, it by no means follows that these men would assume the role for it is unlikely that their personal backgrounds would have fitted them for it.

Therein lies the major difficulty which confronts a developing territory for in the early days it will almost certainly be necessary to import members of the managerial and entrepreneurial class from overseas. Problems associated with this situation have been examined in Chapter III and we need not repeat what has already been said. It is sufficient to notice here that the presence of a foreign-dominated entrepreneurial and managerial class inevitably gives rise to discontent if not to open hostility and the employment of such a group must be viewed as a temporary, though not necessarily a short, phase of development; it is, however, only as development proceeds and as nationals gain experience in business techniques and methods that a domestic class of entrepreneurs is able to develop.

If there is a lesson to be learned from this chapter it is that there is no short and easy way to increased agricultural productivity in the humid tropics. There can be, however, no slackening in the efforts which are now being made to find ways of increasing production with the ultimate aim that the inhabitants can adequately feed themselves and also maintain a higher standard of living in the future than has been possible in the past. The Western industrialised territories, which were once in the situation of the L.D.C.s today, have shown by their own efforts that this can be done; they have indeed been so successful that in 1973, to take a specific example, the developed countries (which support only about 18 per cent of the total world population) accounted for about 65 per cent of the world's total economic product, and the only real way to redress this imbalance is for the Third World to increase substantially its share of total world production. So far in this study we have concentrated mainly upon agricultural development and we have only incidentally mentioned industry; it is to this aspect of development that we should next turn.

Chapter VII

The Background to Industrialisation in the Humid Tropics

STUDIES on the humid tropics have hitherto paid for the most part scant attention to the development of industry in that vast region. Gourou (1968), for instance, deals with the topic in a single short chapter in which he comments upon the feeble development of industry in the whole area and then concentrates almost entirely upon possible industrial uses of the timber and timber products of tropical forests. He points out that there are very few industrial regions in the tropics, the most noteworthy being the Calcutta–Jamshedpur region in India and the Rio de Janeiro–São Paolo–Volta Redonda region of Brazil. While the overall picture has been modified in recent years it has not substantially changed and it is still true that Third World industry accounts for only a very small proportion of total world manufacturing activity; it produces in fact less than one-fifth of the world's manufactured goods. Manufacturing provides less than 10 per cent of the Gross Domestic Product* in the African territories, 15–20 per cent in Asia and 15–30 per cent in Latin America and it generates less than one-third of all employment in those areas except in Hong Kong. The interested enquirer is naturally led to ask why this is so.

THE EARLY STAGES OF INDUSTRIAL DEVELOPMENT

There is perhaps a tendency in some quarters to underestimate the strength of the very real problems involved in the transforming of a non-industrialised to an industrialised territory. After all, this process took a very long time in Western Europe and the U.S.A. though it is difficult to set a precise period since no exact time limits can be recognised. As a preliminary, however, we might observe that the work of Rostow in this field throws considerable light on the whole question of industrial-

* The terms *Gross Domestic Product* and *Gross National Product* are used of attempts to quantify a country's total annual output. The G.D.P. is the value of the gross national output, including exports, but excluding imports; the G.N.P. is the gross national income making no allowance for depreciation of equipment or stock.

isation, though since the work is well known we need here only briefly refer to it. He postulates first a long period, possibly a century or more, when the conditions necessary for economic "take-off" are slowly being established. During this period we can normally trace the development of a stable society within which economic activities gradually become more implanted. Enterprising men appear who are willing to mobilise savings and use the money in risky ventures, frequently in commerce, while commercial markets notably for farm products, domestic handicrafts and imported consumption-goods widen. Financial institutions make their appearance and a rudimentary infrastructure is developed, notably in transport and communications. The rate of productive investment may reach 5 per cent of the Gross National Product though this is likely to keep only slightly ahead of the rate of population increase.

The "take-off" stage usually begins with a sharp stimulus, and this may come either from within the community or from outside. It may come following a political revolution or through a far-reaching technological innovation; sometimes the stimulus may be provided by a developing trading opportunity such as that enjoyed in West Africa by British textile producers in the middle of the eighteenth century or by the opening of British and French markets to Swedish timber a century later; it may develop as a result of a crisis in international trade such as served to stimulate domestic industrial growth in Argentina and Australia during the Second World War when imports of manufactured goods from overseas were rudely interrupted. These are examples of stimuli which in the past have produced significant surges of industrial activity, surges which in their turn developed sufficient momentum to become self-perpetuating and which have therefore taken territories into the third stage of "sustained economic growth."

One or two examples of this kind of development have already been mentioned but it might be useful to examine a little more closely one specific example, that of Britain where the period of initial preparation (Stage 1) had certainly begun by the time of the Civil War—which, indeed, in large measure was itself a result of the structural changes in society involved. The take-off (Stage 2) might be viewed as beginning in the 1780s following the invention of the new machinery which was to revolutionise the cotton textile industry, and it proceeded throughout the rest of the century and through the following hundred years; indeed, in a vastly changed form it continues today. Rostow believes that the stage of industrial "maturity" in Britain was reached near the middle of the last century which means that about three-quarters of a century had elapsed since the "take-off" but we must remember that

Britain enjoyed many early advantages which newly developing territories today do not. Among these advantages we may note the following.

1. Britain was first in the industrial field and had therefore to face no competition for her manufactured products.

2. She possessed very substantial supplies of merchant capital which became available for industry.

3. There was no government interference with industry. Industrialists enjoyed a freedom from legal restraints which permitted a forceful, if sometimes ruthless, advance.

4. Britain at that time benefited from a favourable assemblage of fuel and raw materials. Most of the minerals involved were present in what would now be considered uneconomically small amounts but in those days production was naturally carried on on a very small scale judging by today's standards, and the limited amounts of iron ore, to take one example, were then worth mining. Such deposits gave industry a start upon which it could later build.

5. Although the development of the Industrial Revolution in Britain was very "jerky" and uneven, and by no means as smooth as is sometimes imagined, the free play of market forces which then was allowed to operate meant that industries were able to develop at a "natural" rate as the demand for their specific products grew. Individual mistakes were inevitably made but these were of scant importance except to those directly concerned.* The situation today is quite different when comparatively large-scale individual enterprises are sometimes begun in tropical territories under public auspices frequently for reasons of prestige and without a true regard for the economic situation. This kind of "off-balance" development can seriously hinder overall economic growth as it means that resources are used but not to the best effect.

6. Britain was far more thoroughly prepared for the Industrial Revolution than is often realised, a point which was implied earlier. Manufacture, for instance, had been traditionally a feature of country life. Villages customarily supplied their own clothes, implements, buildings, flour, bread and beer, and only the rich folk of the countryside would send further afield for their furniture, china, clothes and other refinements. Moreover, many villages came to produce not only cheap essentials for general use but also specialised goods for sale in a wider market. Such goods included woollen textiles and also surprisingly sophisticated products such as grandfather clocks;

---

* See, for example, the case of the industrial "false start" examined by the present writer elsewhere (Jarrett, 1974b, 127–130).

indeed, the present writer owns an eighteenth-century grandfather clock which still keeps excellent time and which was made, according to the inscription on the clock face, in the small Worcestershire village of Pebworth by W. Johnson. All this development meant that there was a sometimes unsuspected reservoir of skilled labour and initiative awaiting the onset of the Industrial Revolution.

It is probably fair to say that such a situation does not exist today among rural societies generally in the humid tropics except in a few instances. In south-western Nigeria, for instance, a textile industry has been developed in some centres among the Yoruba inhabitants while other centres such as Oyo are well known for their leather work. The delta-dwellers of Vietnam have developed specialisms in crafts in their villages, each village specialising in one particular craft; the various articles (cloth, pottery, ironwork, basketwork, for instance) are traded in the market towns. Such developments are somewhat unusual today though they were much more usual years ago for it is unfortunately true that one effect of European intervention in the tropical world was the destruction of domestic industries. Indian craft textiles, for instance, found ready markets in eighteenth-century Britain but as the British Industrial Revolution got under way and Lancashire manufacturers sought to secure markets or their produce import duties were levied by the British Parliament to keep out the Indian fabrics and this dealt a severe blow at Indian production. Imports of cheap British textiles into India completed the destruction of the native craft industries. A similar type of protectionism led Portugal to ban the manufacture of all but the coarsest cloth in Brazil between 1785 and 1808 in order to eliminate competition for her own manufactured fabrics. In other cases such as that of the iron workers of Sierra Leone the native industry simply withered away before the onslaught of imported manufactured goods.

It is only fair to say, of course, that developing territories today can enjoy advantages that the Britain of the Industrial Revolution could not; for instance they can draw upon a wealth of knowledge and expertise from other countries which was not available in the early days. The fact remains, however, that despite all these initial advantages it took many years of trial and error, success and failure, affluence and the direst poverty, before Britain emerged as a mature industrial nation and we cannot expect that comparable developments elsewhere can be achieved easily or quickly. Present difficulties, however, form no basis whatever for despondency regarding the prospects for future industrial progress in the humid tropics.

THE PRESENT SITUATION

Of the need for industrialisation in the L.D.C.s of the humid tropics there is general agreement though some writers have questioned its exact value. Bauer and Yamey (1965, 237–8), for instance, criticise the too-facile acceptance of the belief that manufacturing industry offers a panacea for economic stagnation and poverty. They argue that manufacturing industry is simply one type of economic activity and that there is no special reason why it should at any given time serve best to "promote either the most efficient allocation of existing resources or the most rapid growth of resources." They go so far as to state that "it cannot be inferred from the higher level of real income in more industrialised countries that they [the industrialized countries] owe their advantage to the higher degree of industrialisation."

It is difficult to accept this. It can hardly be doubted that as a result of industrialisation very large amounts of resources which otherwise would remain unused are brought into production; these include natural resources (minerals, fuel and power, and even farm products in some areas) and human resources (enterprise, labour and capital)* and this fact alone must result in the achieving of a higher level of production than would otherwise have been the case. It is difficult to believe that if no modern-type industry existed national incomes in the industrial world would be anything like as high as they are at present, though it is quite true as Bauer and Yamey point out that in some cases manufacturing industry may not contribute very much to national income. Industry in such cases may be at an early stage of growth and it may not be very efficient but this kind of situation need not necessarily continue. It is also true that some territories may have no comparative advantages for industrial development though they may have advantages in other directions (for example for agricultural activities of various kinds) and under these circumstances the deliberate establishment of manufacturing industry, perhaps for reasons of prestige or national strategy, can only be a burden upon the rest of the national economy which in some measure will have to subsidise the industrial enterprises concerned.

For territories with basic advantages for industrialisation, however,

* It is interesting to recall that in earlier years in Britain when there were very few outlets for the productive use of capital rich people used their money reserves for consumption expenditure, for example on building and beautifying churches, dwelling houses, ornamental parks and gardens which did nothing further to increase productivity though employment was admittedly provided for a limited number of maintenance staff such as gardeners and domestics.

there can be a clear gain as manufacturing enterprises get under way. Such advantages can take many forms and may stem from the physical or the human environment. Among the former we may list the possession of or proximity to supplies of fuel and power, raw materials, water supplies (modern industry uses very large quantities of water), and a generally helpful physical landscape with, perhaps, a fair amount of low, flattish land for the location of urban centres and factories and to provide natural routeways and harbours; the latter may include the existence of a sufficiently large market together with available supplies of capital, management and labour. These points have been fully discussed elsewhere by the present writer (Jarrett, 1974b).

Even where the prevailing circumstances seem favourable, however, L.D.C.s in the humid tropics which are attempting to establish industrial development for the first time may find the going difficult. One main reason for this is that such territories have to purchase overseas essential equipment and skills and the ability to do this normally depends upon the ability of the territory to borrow capital from foreign sources as it is only under special conditions, for example the sale overseas of large amounts of high-value minerals, that the necessary capital is available from domestic sources. A further difficulty has often arisen because since the close of the Second World War L.D.C.s have often found that the terms of trade* have tended to move against them and they therefore have received lower unit prices for their exports with the passing years. This situation is made even more difficult if they have to pay higher unit prices for their imports. Crawford (1964) has examined the reasons for this development and the following are the main points which he makes.

1. In developed territories industries tend to use smaller amounts of raw materials to produce any given output of manufactured goods as time goes on because of technological improvements. Demand for raw materials correspondingly slackens.

2. As incomes in the developed world rise the demand for primary products (including foodstuffs) does not rise proportionately, largely because of increasing expenditure in the tertiary sector—on holidays, recreation and education for instance.

3. Export of foodstuffs from the L.D.C.s is also restricted by various forms of agricultural protection established in developed

---

*The *Terms of Trade* represent the relationship between export and import prices. Import prices are expressed as a percentage of export prices. If the prices of imports rise more than those of exports the terms of trade are said to be adverse or to be deteriorating.

THE BACKGROUND TO INDUSTRIALISATION 165

countries which feel it necessary to safeguard their own farming interests for economic and also strategic reasons.

4. Some raw materials, textiles and rubber for example, have to meet increasing competition from synthetic substitutes developed and manufactured in the industrialised countries.

5. Increasing production of primary products and an expanding volume of exports of these goods have between them lowered unit prices.

We must, however, remember that the world economic situation has changed in many respects since Crawford wrote and some modification of his general argument is in some respects necessary. For example, with increasing world demand the prices of foodstuffs and raw materials have shown marked signs of stiffening in recent years and the situation facing the L.D.C.s is therefore more favourable today than it was in 1964. It is also true that quite substantial sums which are spent by inhabitants of the developed territories on holidays (*see* 2 above, tertiary sector) are finding their way to L.D.C.s which offer tourist attractions; various countries in the Caribbean and parts of Africa are cases in point.

AGRICULTURE AND INDUSTRY

It is perhaps desirable at this point to emphasise that there should be no fundamental dichotomy between agriculture and industry in the humid tropics or elsewhere. Not only do both spheres of economic activity contribute to the well-being of the inhabitants of the territories concerned but they closely interact with each other, for not only do farming communities produce food for those engaged in the secondary and tertiary sectors but at the same time much industrial activity is directed towards agriculture. Fertilisers, pesticides, herbicides and machinery of various kinds are all produced by industry for use on the farms, while transport routes are developed to serve farming areas and research is applied to farming problems. The interaction between the various economic sectors can be much closer than is often realised.

Figure 21 attempts to show the development of these relationships over a period of time along what has been termed the *development continuum*. To a large extent the graph is self-explanatory. It illustrates the case of an expanding population which changes from a completely rural farming base to a pattern in which farming, manufacturing, mining and service industries are all represented. The farming population does not of course increase anything like as much as the total population and

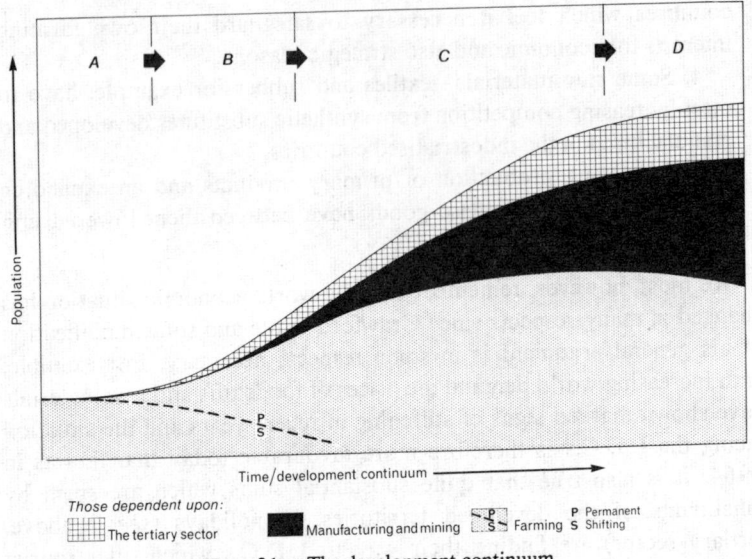

FIG. 21.—The development continuum.
For discussion *see* text.

it may actually decline at a late stage of development. This can come about as farming becomes more efficient with the help of better techniques; the non-farming population is absorbed into the mining, manufacturing and service sectors of the developing economy.

Four stages in the continuum are recognised, each one gradually merging into the subsequent one. The stages are as follows.

A. This is the traditional stage of subsistence cultivation which is dependent upon labour rather than capital of which there is very little for productive purposes.

B. We see here the beginning of the growth of the tertiary sector as administrators and the pioneers of the professional classes (lawyers, doctors, traders and teachers, for example) make their appearance. A start may be made with mining if conditions are appropriate and with manufacturing. It is still in farming, however, that the bulk of the population is engaged though because of the increasing numbers of people some modification of the traditional pattern may become necessary. This may come about with the development of permanent field cultivation with a corresponding decline in shifting cultivation as is suggested by the pecked line in the diagram, though farming is still dependent upon labour rather than capital.

C. In this lengthier stage the levels of both secondary and tertiary activities rise sharply while that of farming proportionately decreases. The numbers of those directly involved in farming are likely only marginally to rise and they may even begin to decrease near the end of this stage of development. During this period the primacy of population increase as the main agent of change gives place to that of technology as the interaction between industry, the tertiary sector and farming strengthens while the importance of labour as a factor of production proportionately declines and that of capital increases.

D. Finally, agriculture becomes even more strongly organised along commercial lines; the special case of plantation agriculture which was examined in Chapter V illustrates this though commercial agriculture is by no means limited to plantation enterprise. The main accent is now upon capital, in farming, mining, manufacture, and even in the tertiary sector (the contemporary widespread use of computers offers a dramatic example of this). Proportionate and actual numbers of farm workers decrease though output continues to expand.

There is not meant to be in the above analysis any suggestion that the exact pattern illustrated in Fig. 21 and described above is followed in every case. Land reform, for instance, which has been discussed in Chapter VI, may be undertaken under political pressure at any time during Stages *A* or *B*, but as was emphasised earlier the timing must be right if it is to bring real and fairly rapid benefit to the community. Study of the pattern illustrated in Fig. 21 suggests that reform during Stage *A* might well be premature as the essential elements necessary for more efficient agricultural production are absent; on the other hand if such reform is delayed into Stage *C* development may well be retarded. Probably about mid-way through Stage *B* is theoretically the optimum time. The continuum also suggests that attempts made to stimulate agricultural efficiency during Stage *A* are likely to be useless, again because the other essential elements are absent. This conclusion accords well with previous observations made in Chapter IV that attempts to increase the efficiency of shifting cultivation, for example by increasing the work force or by employing animal power, generally seem to fail. On the other hand improvements can confidently be expected as Stage *B* draws to a close and Stage *C* is reached, as permanent field agriculture strengthens and shifting cultivation declines, and as industrial feedback to farming gathers force. A correlated point, not perhaps particularly welcome in this age of instant solutions and panaceas, is that economic development cannot successfully be forced out of context. While it might be possible in some instances slightly to speed up development along the continuum it is not normally possible to omit stages and to superimpose, say, late Stage *C* or early Stage *D* upon Stage *A* or early Stage *B*. Any attempts to do this are likely to result in waste and sad disappointment.

THE ECONOMIC BASES OF INDUSTRIALISATION

We are now in a position to examine more closely the economic bases of industrial development in the humid tropics, and this can best be done by considering in more detail the role of capital, for it is the increasing use of this factor which distinguishes developed communities from traditional ones. It is capital which becomes the vital factor as development proceeds and it has already been emphasised that it is the availability of capital which makes it possible to move from low to high levels of productivity. The problem of capital formation is therefore a vital one.

Capital Formation

The central dilemma of the L.D.C. is that it can provide only very limited supplies of capital from its own resources because the *per capita* income is low. This in turn means that productivity and therefore incomes remain low. This "vicious circle of poverty" was recognised many years ago by Nurkse (1952) and it is expressed in diagrammatic form in Fig. 22(*a*). The very poverty of a poorly developed territory is itself a fair guarantee of continued poverty.

This is one main reason why for so long many poor countries have remained poor; they are in the grip of an economic situation which is very difficult indeed to change. This is the static economy to which reference has earlier been made. It is true that as a result of increased diligence and more efficiency some increase in productivity may be expected even under these circumstances, but the increase will be limited; the vicious circle will still operate albeit at a slightly higher level. A different economic structure, however, is needed before the situation can radically be transformed and this can come about only as a result of increasing capital investment which will make possible the movement along the development continuum from Stage *A* through Stages *B* and *C* (Fig. 21). Mountjoy, writing in 1966, believed that

(a) The vicious circle of poverty.

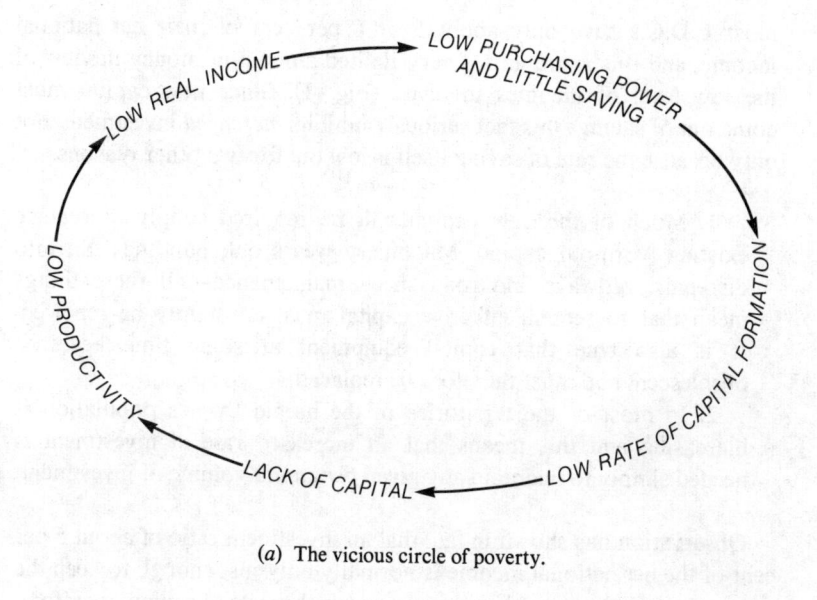

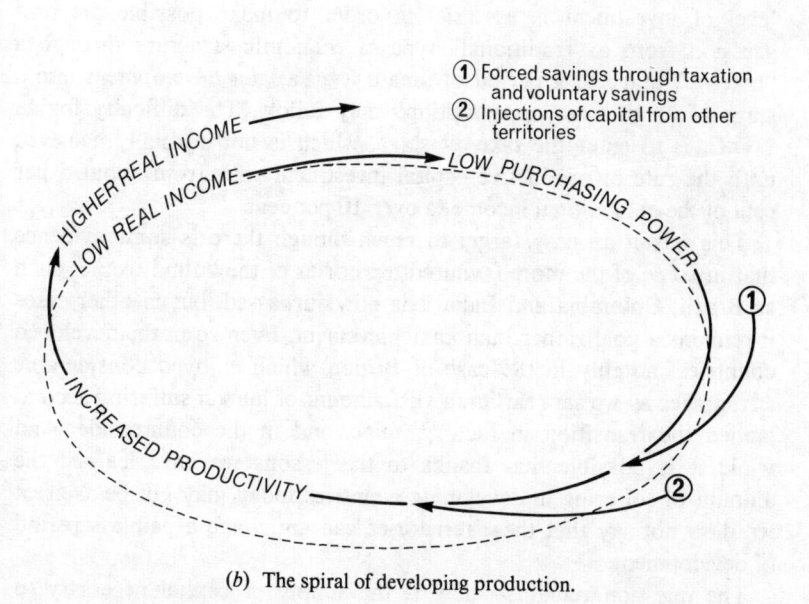

LOW REAL INCOME → LOW PURCHASING POWER AND LITTLE SAVING → LOW RATE OF CAPITAL FORMATION → LACK OF CAPITAL → LOW PRODUCTIVITY → LOW REAL INCOME

(b) The spiral of developing production.

LOW REAL INCOME → LOW PURCHASING POWER → ① Forced savings through taxation and voluntary savings ② Injections of capital from other territories → INCREASED PRODUCTIVITY → HIGHER REAL INCOME

Fig. 22.—Capital formation and income in developing countries.

most L.D.C.s save only about 5 or 6 per cent of their net national income, and this represents a very limited amount of money in view of the low level of incomes involved (Fig. 1). Since new capital must come out of savings this fact seriously inhibits *increased* investment, not only because the rate of saving itself is low but for two other reasons.

1. Much of the new capital will be required simply to replace existing worn-out capital. Machinery wears out, buildings fall into disrepair, railways and roads need maintenance—all these things mean that to remain effective capital must constantly be renewed. It is also true that capital equipment at some time becomes obsolescent and must therefore be replaced.

2. In most of the territories of the humid tropics population is increasing and this means that an increased rate of investment is needed simply to maintain any given *per capita* volume of investment.

Observation has shown in fact that an investment ratio of about 5 per cent of the net national income is normally only just enough to keep the *per capita* investment ratio constant, and at this rate of saving, therefore, it is possible only to maintain the existing economic status. If any significant increase in productivity is to be achieved a notably higher level of investment is essential in order to make possible the progression from a "traditional" type of economic structure through a "take-off" stage to a period of "maturity;" a later development into a stage of "high mass consumption" may follow. The difficulty for an L.D.C. is to enter the take-off stage, which is not normally achieved until the rate of productive capital investment rises from about 5 per cent of the net national income to over 10 per cent.

This is not an easy target to reach though there is some evidence that in some of the more favoured territories of the humid tropics such as Brazil, Colombia and India it is now surpassed, but in other cases it remains a goal rather than an achievement. Even so in the developed countries, notably in the case of Britain which enjoyed considerable advantages as we saw earlier, a vast amount of human suffering accompanied the transition in factory, mine, and in the countryside. And while it is possible that thanks to the lessons we have learned the amount of suffering in developing countries today may not be as great we dare not say that these territories can anticipate a painless period of development.

The question remains—how is the supply of capital necessary to ensure progression into the take-off stage to be secured? This is not an easy thing to accept but it can be argued that even a poor country can save 10 per cent of its net national income *if there is the will to do it*. This does not mean of course that a 10 per cent rate of saving could be expected from everyone; that would be manifestly absurd. But it does mean that the affluent members of the community must be prepared to forgo present pleasures in the interests of future prosperity. In the L.D.C.s it is not unusual to find extreme inequalities of wealth which are far in excess of anything known in the developed countries where taxation has done much to level out effective incomes, and the pressing need is to direct the wealth which undoubtedly exists into productive channels. This is not easy for two main reasons; first, there is no tradition of financial enterprise in the humid tropics, and, secondly, there are few financial institutions to act as intermediaries between savers and entrepreneurs, even if these exist. There is, however, the possibility of taxation and the subsequent initiation of government-sponsored projects, and the possible effect of this course of action is shown by arrow (1) in Fig. 22(*b*). As savings induced by taxation in addition to voluntary savings are put to productive use, the vicious circle of poverty can begin to give place to the spiral of developing production.

A second possibility remains, that of securing supplies of capital from more developed countries in the form of loans or outright gifts. This is the point of arrow (2) in Fig. 22(*b*). Unfortunately foreign capital is frequently viewed with considerable disfavour, perhaps for three main reasons.

1. The fact that a person or a country may recognise an urgent need to seek help in the form of grants or loans does not in itself reconcile the recipient to the situation which arises when help is given. No one likes to feel beholden to another and a person or a community with pride is in fact likely to resent the whole set of circumstances which makes the receiving of aid necessary.

2. There is always the fear that economic dependence may lead to political dependence and a loss of sovereignty. This is in large measure the fear which leads to much contemporary agitation against "neo-colonialism."

3. In years past most private investment in the lands of the humid tropics took the form of investment in extractive industries and to a lesser degree in agricultural activities such as the establishment of plantations. In those days possibilities for investment of profits in other sectors of the economy were non-existent and the only thing an investor could do with his profits was to take them out of the country for possible investment elsewhere (after the distribution of dividends)

where opportunities presented themselves. The benefit to the borrower country under these circumstances was therefore minimal and it was natural enough that resentment should build up against the foreigner who apparently simply lent his money in order to enrich himself while rendering minimal aid to the borrower. In the changed situation of today opportunities for re-investment of profits are opening up in the countries of the humid tropics, but old feelings die hard and it is paradoxical that some countries which desperately need foreign capital create circumstances which frighten it away. Fear of civil disorder or expropriation will always discourage overseas investors but so will legal restrictions on the repatriation of profits or of capital. The same is true of any form of discriminatory taxation, of legislation requiring foreign investors to employ a certain proportion of local labour, of directed investment and of the imposition of overseas trading licences which can seriously impair efficient working; yet these are all examples of actual restrictions which at different times have been imposed by territories seeking to borrow capital from overseas.*

CAPITAL AND PRODUCTIVITY

We have today moved a very long way from the older geographical viewpoint that it is the possession of natural resources which is the main factor which determines the development and prosperity of a country. Such a proposition is in any case demonstrably untrue; the United Kingdom for instance has never been well endowed with natural resources except in coal reserves. Even its soils for the most part were naturally of indifferent or mediocre fertility and it is the efforts of its farmers which have improved them out of all recognition. It still remains to be seen how extensive are its reserves of mineral oil though they appear to be considerable, but the value of this source of wealth lies in the future and not in the past.

This example of the United Kingdom demonstrates that there is much more to prosperity than the simple possession of natural resources; what in fact is critical is not the relation between natural resources and wealth but between capital and output. Gross National Product, so many authorities would argue, depends essentially upon two factors: the amount of capital productively used within a

---

*This whole question of capital formation and acquisition is examined more thoroughly in a sister book in this series: (Jarrett 1974b, 302–9).

community and the relation between that capital and output (the capital:output ratio). This formula may be applied to the economy as a whole or to individual sectors and it is important to remember that for the whole economy any increase of overall output by a single unit normally requires an input of three or four times as much capital; the capital:output ratio, in other words, is 3:1 or 4:1. In some sectors, however, for instance in heavy industry, a considerable amount of overhead capital (for example buildings, machinery, power supplies and other essential infrastructure) is essential and this results in a comparatively high ratio which may be as high as 8:1. Investment in agriculture on the other hand frequently yields more attractive returns, sometimes as much as 1:1. It is very important therefore for investment, particularly in the L.D.C.s, to be directed into the most favourable channels and it is increasingly realised that investment in heavy industry can easily be premature and can yield very disappointing results, as has been the case in India and Pakistan. In the long run, with the fuller utilisation of agricultural, industrial and labour resources which comes with more advanced development capital:output ratios may be dramatically reduced.

THE BALANCE OF DEVELOPMENT

An issue of the first importance in all economies is that of sectional balance, and this is of particular importance in the territories of the humid tropics because if a wrong balance is struck, say, between agriculture and industry, or even within the industrial sector itself, the economy as a whole is bound to suffer. In the more spacious days when the present developed territories were themselves at an early stage of development economic enterprises tended to establish themselves as the need for them arose and a kind of natural balance resulted, but in the humid tropics today development rather tends to take place as a result of central planning decisions and the question of maintaining a correct balance is therefore a very pressing one indeed. There is no doubt at all that some economies have suffered because of ill-conceived enterprises started up for reasons of prestige, but it is possible that as time goes on and as the damage which such enterprises can do to the economy is more fully realised the number of such enterprises will decline.

The concept of "balanced growth" is attractive and apparently more than reasonable and many writers support it (*e.g.* Nurkse, 1971), but as in the cases of so many superficially attractive arguments it bristles with difficulties which become increasingly apparent upon closer examination. The optimum economic situation in territories like Brunei or Mauritius on the one hand must be of a completely different kind from that appropriate to India and Brazil on the other, and while these examples are chosen from the extreme ends of the tropical spectrum of countries similar differences must arise between those closer to the centre—but in these cases the differences may not always be so obvious.

Apart from these territorial differences, however, the question arises whether it is possible or even desirable to aim from the outset at balanced economic growth. Rostow (1971) has shown that in the earlier developmental years of present industrialised territories the take-off point was reached through the powerful development of one powerful "initiator" or "trigger" industry working in a favourable economic climate. In Britain it was the textile industry; in the United States, Germany and Russia it was railway construction; in Sweden it was the timber industry (which was later further stimulated by the increasing demand for pulp) based on the steam saw; and in Denmark it was the shift to meat and dairy products after 1873. It is clear that no single sectoral activity or sequence can be in every case responsible for economic take-off; it varies, as we have previously argued, according to circumstances. When once established, however, an economic leader can lift an economy into self-sustaining growth though at least three basic requirements must be met.

1. There must be a developing and effective demand for the product (or products) of the leading (initiator) industry. This may come about as wealth is released from hoarding or from non-productive consumption uses (a point made earlier in the footnote, p. 163) and put into productive investment which in its turn generates additional incomes which confer additional purchasing power,* or it may be helped by imports of capital or by a general increase in incomes. It is, of course, likely that more than one of these factors will operate together to increase incomes and therefore stimulate market demand.

2. There must be in the leading sectors a very considerable expansion of productive capacity together with similar expansion in associated industries. Thus, the development of the British textile

---

*Any money capital which is spent on investment goods (producer goods) is received as income by the sellers of those goods; some of this income goes to the management, some to shareholders and some to the workers concerned in the manufacture of the capital equipment. It is the incomes thus generated which produce the increased purchasing power referred to.

industry stimulated associated developments, for instance in engineering (largely through the design and manufacture of the new textile machinery), in the coal, iron and steel industries (to provide fuel and power together with the essential raw materials for the machinery), and in transport (to assemble raw materials and distribute the finished products to markets and ports); this in turn further stimulated engineering and the production of fuel and raw materials as sleepers and rails must be made and laid down as permanent way, and locomotives and rolling stock must be built, to mention just some essential developments. In other words, a chain reaction within the economic body must be set in motion so that eventually all sectors of production benefit. This "chain" will include agriculture as the new *corps* of industrial workers, better paid than ever before, will initiate a market demand for greater quantities of more varied foodstuffs than has been the case hitherto.

3. There must be the will and the ability to accumulate large amounts of capital to spark off and maintain development in the leading and the associated sectors. This presupposes some restriction in consumer spending, a condition which has been achieved in earlier years by the payment of comparatively low wages though this is less likely today. Furthermore, the capital must be ploughed back into the developing sectors at a considerable rate, but this requirement is clearly very difficult to satisfy in the face of organised demands for high wages which are for the most part dissipated in consumption spending, and of penal taxation on profits which is used largely in social spending, most of which again is of a consumption nature.

In a sense, of course, the argument which we have so far followed does imply that a kind of balanced growth is essential for successful economic take-off and further development; the chain reaction referred to as essential may be interpreted as itself constituting a kind of balance. Nurkse (1971) recognises this fact clearly when he points out that a leading sector must be one with a broad economic base, and he quotes the shoe industry as a possible negative example. Shoe producers operate within a much narrower market than do producers in leading industries such as those mentioned above, yet clearly they depend for success upon sales of shoes. Alone, however, they can do little to ensure increased sales; if nothing happens in the rest of the economy to increase incomes people will not have extra money to spend on shoes and the market therefore is likely to be too small to sustain a developing shoe industry. This is particularly likely to be the case in L.D.C.s where the inhabitants are hard enough pressed anyway to purchase absolute necessities without renouncing any of these in order to buy additional pairs of shoes. The new industry under these circumstances is likely to prove a failure though it could be very successful in a more developed economy where spending power is greater. The essential point at issue is that a narrowly based industry like shoe manufacture can never of itself be an economic "leader" as can a broadly based activity like the textile industry.

A related problem which is of great importance to the territories of the humid tropics is that it is now possible, with the aid of modern technology, to satisfy domestic demand for many manufactured goods with surprisingly few up-to-date and efficient industrial establishments which employ a comparatively small labour force. Mountjoy (1971, 210) states, for example, that a single modern factory which employed roughly fifty workers could produce enough biscuits to meet all market demands in Ghana and Nigeria together. Clearly such enterprises cannot act as "leaders" of industrial development either though they may well play a small part in helping to make possible a more general economic advance.

One thing which the term "balanced growth" cannot mean in a developing economy is that all sectors advance *pari passu*, for there could not be sufficient capital generated in the early stages at least to make such a development possible. If, in an attempt to bring about such a general development, potentially thriving sectors are mulcted of much of their capital through taxation, the proceeds being used to encourage growth in other sectors, the result must be that the potentially thriving sectors, starved of their own capital, will wither while the thinly-spread resources are not likely to be sufficient to stimulate growth in other sectors. Under these circumstances any general economic advance is likely to be halted, or at least slowed down very considerably.

SOME OTHER PROBLEMS OF INDUSTRIAL DEVELOPMENT

Even when industrial development has begun in an hitherto undeveloped territory we cannot expect that output will rise automatically and proportionately to any increase in investment "in accordance with some magical capital:output ratio" (Reddaway, 1971). Output depends upon a great deal more than simply capital, a point which has previously been stressed, and it is fatally easy for theorisers from developed countries to assume the existence of the other essentials though the likelihood is that they are lacking, or if they do exist it is not in amounts needed to match the growing need for them. If the reader refers back at this point to the section dealing with the failure of the East African Groundnut Scheme in Chapter VI, for example, he will review an actual case which illustrates many of the main points. For instance, labour must be available to make possible any extension of economic activity and the term "labour" includes skilled labour and management as well as unskilled workers, while raw materials, component parts, fuel and power and transport facilities are all essential but these are requirements which may at times be more difficult to supply in the necessary quantities and qualities even than the capital equipment necessary for an enterprise. Some shortages, including that of skilled labour, may be met with the help of imports if the balance of payments situation permits but this expedient is not likely to afford a complete solution to the problem. Power resources are particularly vital, for a shortage of power can shatter an entire development programme while it is also essential that an adequate infrastructure be provided for if it is not capital investment in other sectors will yield very disappointing results. In general we must recognise that production will be held back if these other vital prerequisites are not available because bottlenecks will hinder the free flow of raw materials, partially manufactured and completed goods. Factories will therefore be working at well below their potential capacity while the distribution of finished products will also be inefficient and waste will occur.

It is, in fact, not infrequently necessary in particular cases to invest to a fairly large extent considerably in advance of present requirements though it may be hoped that with increasing development more and more of the "slack" will come into use. This situation arises because some types of investment are extremely "lumpy." In such cases the investment cannot proceed along a fairly even course but has to take place in large "lumps" or not at all. The example usually given of this is that of a railway, a useful one because the essential point at issue can be readily appreciated. Trade of any kind is not possible without movement and for a very long time the most useful form of transport was the railway. A railway, however, is a very expensive and large-scale piece of capital equipment yet it is not normally possible to start it on a small scale and build it up as demand for its services increases. A railway designed to transport goods between two places must *in the first place* cover the whole distance between them; nothing less will do. And the railway must include all the essential features of any railway—sidings, signal boxes, stations and so on—to be operable. This means that very considerable capital investment must go into the construction of the railway though the completed rail system may be too elaborate and effective for present requirements. There is of course always the hope that traffic will subsequently grow to the point where the full potential of the railway can be utilised. As we shall see in the next chapter, however, there is good reason to doubt today whether railway investment is any longer very useful except in special cases—for example to move large quantities of a heavy and bulky mineral from a mine to a port.

We have earlier made the point that certain basic requirements must be met on the supply side if an L.D.C. is to reach and pass the point of economic take-off. It is equally true, as we have already implied, that market conditions must be favourable. It is no use producing goods for sale if people cannot or will not purchase them as we saw in our example of the shoes where one of the points at issue is that for economic success there must be a sufficiently strong market demand for the products of the newly established industries to justify full and efficient working of the various producing units. The existence of this demand will depend upon the generation of a broadly based income structure which will confer increasing purchasing power on the community as a whole—or at least upon large sections of it.

A closely related point is that there must also come into existence a marketing structure which will permit manufacturers to distribute their goods widely and which also includes a large number of retail establishments where purchasers can easily obtain the goods which they are willing and able to buy. Such a structure will include not only the "mechanical" means necessary to achieve this end (railways, roads, shops and other component parts of the infrastructure) but also credit facilities designed especially for retailers and the existence of law and order within the community without which all schemes for economic advancement and maintenance must collapse. Indeed, Galbraith (in *The Affluent Society*) goes so far as to suggest that law and order themselves constitute a basic requirement for economic production which would become impossible on anything but the smallest scale if theft and wanton destruction of property and goods were rife.

THE ROLES OF AGRICULTURE AND INDUSTRY

We are by now familiar with the discussion regarding the relative importance of the roles played by agriculture and industry in stimulating economic progress in the early stages of a country's development.

On the one hand there is the argument that increased agricultural production which stems from more efficient farming practices is an essential pre-requisite for a successful industrial revolution, and those who hold this view can justifiably point to the experience of countries which have successfully passed the take-off point. The case of Britain is well known in this respect and it is possible to trace similar sequences elsewhere, for example in Mexico. This line of thought takes into account the fact that traditional agriculture is not capitalised and those who practise it will normally see no need to change their customary methods. Indeed, even if they wished to do so it is difficult to see how they could manage it. There is under the prevailing circumstances no commercial incentive to change from within the farming community itself while there is little or no external incentive to potential investors to interest themselves in agriculture. There is therefore little likelihood of developments on such a scale as to make possible any economic take-off within the farming sector until we see the emergence of an enlightened and progressive majority of farmers who are ready and willing to exploit the possibilities of more efficient farming techniques, improved land tenure, improved transport facilities and more effective forms of marketing. Those who see in such developments essential pre-requisites for economic take-off will argue that it is such factors which lead to the emergence of a well-nourished farming population with the increased purchasing power necessary to continued and sustained economic progression, but it is not easy to envisage such a situation developing in the lands of the humid tropics within the foreseeable future; there are too many uncertainties inherent in it.

There are equally those who would argue that the above sequence of events could not possibly take place without some external stimulus and these would accord the major role in developmental economics to industry because, they argue, only industrial development can successfully break the vicious circle of poverty and induce economic take-off. Protagonists of this view argue that standards of living are highest in industrialised countries where the value of output per worker is higher than elsewhere. Industrial growth is also said to be cumulative and can stimulate development throughout the economy and this should bring about the greater economic diversity which most territories in the humid tropics at present lack. Furthermore manufacturing industries are more flexible and adaptable than is agriculture and they can adjust themselves more quickly to changes in demand. This is important as the demand for most manufactured goods is more elastic than it is for agricultural products, and it comes about because the proportion of

fixed to operating costs is much lower in industry than in agriculture.*
This means that industrial production can more easily be increased or
decreased in response to changes in demand, a feature which is also
made possible by the comparative natures of agricultural and industrial
production for it is not easy, indeed, it may be impossible, quickly to
reverse an agricultural programme after its inception. If ground is
ploughed and planted with a particular crop, for example, the same
ground cannot quickly be used for grazing purposes. A further point is
that although the farmer strives to plan well ahead and indeed must do
so (preparations for any given form of land use frequently begin a year
in advance) he is greatly dependent upon unforeseeable elements of
nature such as weather and pests as the industrialist is not, and he
therefore has much less control over his output than the manufacturer.

WHAT IS THE ANSWER?

It is more than likely that there is no universal answer to the problem
of balanced development because we are dealing with variables, both
human and physical. One chief human variable has to do with markets
and population, to take a particular example. There is clearly a relationship between the two because a market cannot exist without a population, but the size of a market is not necessarily directly linked to the size
of any given population. It has to do primarily with the proportion of
wage-earners in the community and with the general level of incomes
so that a small population with a high proportion of wage-earners and a
high general level of income may well form a larger market than a larger
population with few wage-earners and a low income level. It is fair to
say that in the short run developmental opportunities are limited by
the size of the market. But this in itself is not enough, for we must take
into account the potential size of the market and in this respect population is a vital factor simply because it sets a ceiling upon market
potential. Thus a small country with a small population will inevitably

* Fixed (or supplementary) costs are those which do not vary with output. They are
sometimes known as overhead costs. For instance, the permanent way of a railway must
be maintained in good order whether one train or 100 trains pass along it each
day. Rents and rates do not vary in accordance with the amount of work undertaken
by a firm. On the other hand operating (prime or variable) costs vary directly with
output. Labour, for instance, is normally a variable cost since workers can be taken on
or laid off according to the amount of work to be done, while purchases of raw
materials vary proportionately to business orders. Fixed costs in agriculture are
normally high because of the high cost of the land itself and the care necessary to
keep it fertile and productive. Farm buildings and equipment also have to be kept in
good order irrespectively of the amount which they are used.

THE BACKGROUND TO INDUSTRIALISATION

enjoy a much smaller market potential than a large country which is fairly heavily populated, and the scale and scope of its development will be affected accordingly. Hence, the small country will not normally be able to consider, for instance, a forward economic thrust based upon the development of large-scale industry and it may need rather to concentrate upon building up its agriculture and small-scale manufacturing enterprises. Such a country may be fortunate enough to possess substantial supplies of minerals which it can offer for sale on world markets and these may prove to be a very strong developmental factor.

An outstanding example of a small mineral-producing state is Brunei Dares Salaam with its capital, Bandar Seri Begawan (formerly Brunei Town), which covers a total area of only 5700 km^2 (2200 sq. miles) and which has a total population of only 150,000 inhabitants. This small territory earns a very large annual income from the sale of mineral oil and natural gas (the greater part of the receipts of $687m from taxation in 1974 came from these sources) and because of this there is no income tax; there are, however, free education and health services together with pensions for the old and sick while television and radio services are also provided free by the Government. In addition, the State Air Line is heavily subsidised. Under these unusually favourable circumstances economic advance is at present planned in both the agricultural and industrial sectors. It is hoped that large-scale commercial *padi* growing schemes will produce almost all the country's rice needs instead of the present 20–25 per cent while it is also hoped that overseas companies will participate in large-scale production of oil palms, cocoa and castor oil and that they will also be associated with the establishment of factories to process the resulting produce. Timber will be extracted from the virtually untouched forests for processing in a plywood and veneer factory while a number of small and medium-scale industries may be established; any large-scale industry, however, must be dependent upon a growing export market. In the tertiary sector tourism is being actively encouraged.

While few, if any, other territories in the humid tropics can match the happy situation of Brunei there are some which are comparatively fortunate; these include Liberia with its substantial exports of iron ore, Sierra Leone (diamonds and iron ore) and Venezuela (mineral oil and iron ore). Mining industries in such cases can form development leaders of great significance to the territories concerned.

The case of larger, more populous, territories is of course very different. Here we are dealing with countries with large potential domestic markets and the establishment of large-scale industry in such cases may well induce economic take-off if the other essential circumstances which have been discussed above are favourable. This development is bound to be hastened by the undoubted prestige which is attached to industrialisation in the minds of people everywhere, including many in the humid tropics who rightly or wrongly see in the establishment of industry a passport to prosperity.

THE PATTERN OF INDUSTRY

Because, as we have emphasised, physical and human environments vary so greatly from one country to another economic responses will also vary widely. Even so, some general principles regarding industrialisation in the humid tropics can be formulated. Some industries, for instance, are probably not very suited to the economies of emergent tropical territories from their very nature except in special cases. Heavy industry normally falls into this category, largely because of the very large amounts of capital and the varied skills which are required, because of the unfavourable capital:output ratio, and because of the need for large-scale working to secure efficiency of operation. In the case of the larger territories, however, there may be a case for establishing such an industry as in fact Brazil, Venezuela and India have done.

On the other hand there are many forms of industry which are well suited to establishment in the emergent lands of the humid tropics, even at quite an early stage of development. It is possible to recognise at least six main groups though the following list is by no means meant to be viewed as exhaustive.

1. Those processing materials of local origin for which there is a strong demand in world markets; these include the extraction and refining of vegetable oils, cotton ginning, plywood and veneer manufacture and tobacco processing.

2. Those which utilise materials of local origin in the manufacture of goods for which there is a buoyant domestic demand; these include the production of such commodities as soap, textiles, cement, foodstuffs and beverages.

3. Those which make use of imported raw materials to form the basis of processing industries; examples include flour milling from imported grain, and oil refining. This form of industry is commonly located at or near the ports of entry.

4. Those which use imported component parts in the assembly of comparatively sophisticated end products such as bicycles and automobiles.

THE BACKGROUND TO INDUSTRIALISATION 183

5. Those based upon minerals, either of local origin or imported. Such industries as these, however, can be developed only with reference to the available market. For instance, Sierra Leone and Liberia as we have previously noted have for many years been exporting large amounts of iron ore but the small sizes of these territories virtually precludes the establishment of a domestic iron and steel industry. Venezuela, on the other hand, a larger country with a broader resource base, has for some years been operating an iron and steel industry based on export as well as domestic markets. More details regarding this are given in Chapter VIII.

6. Those which enjoy a "natural tariff" type of protection because of the nature of the industry concerned. In this connection the role of transport costs is paramount because it is these costs which produce the natural tariff. Costs of this nature are of particular importance in industries in which bulky, heavy but low-value raw materials are needed and are available locally. It is also operative where the finished manufactured goods are bulkier and therefore more expensive to transport than the raw materials concerned. Industries which fall into one or other of these categories include the manufacture of cement, bricks and tiles, furniture, and the final assembly of motor vehicles.

Some industries of course will fall into more than one of the categories mentioned above, as the case of cement manufacture will show. Another example is oil refining, which falls into categories 3, 5 and possibly 2. This is an industry which has been established in several territories in the humid tropics and it can form the basis for the establishment of a petro-chemical industry as Nigeria is demonstrating. Such an industry can prove a most valuable asset to a country which needs large amounts of fertiliser to improve its agricultural output.

Chapter VIII

Transport, Markets and Trade in the Humid Tropics

As a territory moves away from a purely subsistence form of livelihood the importance of transport, markets and trade to its economy becomes increasingly important and we deal in this chapter with these three vitally important topics for they form an inevitable accompaniment to any form of economic development. No hierarchical importance is implied by the order in which the three topics are discussed in this chapter; they interact with each other and are strongly inter-dependent.

TRANSPORT

There is one sense, despite what was said in the preceding section, in which transport may be viewed as fundamental, for without transport neither markets nor trade could exist at all. It is many years since King Leopold of the Belgians made his famous statement that *"coloniser, c'est transporter"* and since Lord Lugard (1922) wrote that "the material development of Africa may be summed up in one word—transport" but these statements forcefully remind us that without transport there cannot in fact be any real progress. The theme has by no means been forgotten in more recent years as Clark and Haswell (1970, 191) remind us: "fertilisers, improved strains of seed, education and other objects are all of the greatest importance. But the need for transport is prior to all of these." It is therefore unfortunate that the developing world still lags behind so seriously in transport facilities. The developed parts of the world, for instance, which account for just under 30 per cent of the world's total population, account for no less than 88 per cent of all rail traffic and 78 per cent of the world's lorries and buses (Hilling, 1973, 32). Africa and Latin America together, accounting for 37 per cent of the world's land area, have only 7 per cent of the world's surfaced roads, and even if the enormous areas of Asia are added the total rises only to 23 per cent.

Imagine, as an instance of the situation which would prevail generally in any transportless region, the case of a village which has no contact with its neighbours or with the outside world generally. No goods would be able to enter the village from outside and the villagers would therefore be condemned to an unmitigated subsistence mode of life. Neither could they despatch any goods out of the village so that any form of specialisation beyond that needed for the life of that particular settlement would be out of the question. There could be no possibility of exploiting local resources, for instance specialised forms of crop production or mineral reserves, for any wider use while new ideas and techniques, together with the means of putting these into practice, would pass the isolated community by. The village would remain totally in the grip of tradition and traditional practices.

An interesting and very important corollary of all this is that without transport facilities town life would be quite impossible for towns, by their very nature, must receive substantial inflows of such commodities as food, drink, fuel and raw materials; even in comparatively unsophisticated societies town dwellers must have clothes and buildings for which raw materials are needed. Ports, industrial and administrative centres, market and university towns—none of these would exist, and society, if such a term could be used to describe the ensuing fragmentation of community life, would consist simply of separate and distinct village groups existing in isolation from each other. This is in fact the kind of situation which from time to time has existed not only in fully traditional societies but also along the pioneer fringes of civilisation and also under conditions of civil disunity, strife and lawlessness such as developed for example in Europe after the collapse of the *pax Romana*. Clark and Haswell (1970, 183) remark that something approaching this isolation exists even today in parts of Malawi where transport costs are so high as to be prohibitive and this effectively compels remoter villages to live entirely off their own resources.

On the whole, however, this kind of situation is unusual for even under the most primitive conditions men can walk along paths and tracks and carry goods from village to village unless there is a complete absence of law and order in the countryside. The system known as *porterage* is typical of many parts of the humid tropics even today for the essential requirements are modest; healthy carriers (porters) and bush paths along which they can travel. To leave the matter there, however, would be deceptive for it could imply that human porterage is a simple means of carrying goods and is therefore cheap, whereas in fact it is easily the most expensive form of carrier transport known to man, and the reasons for this are not difficult to find. In the first place, strong and healthy men to act as porters are not very numerous in most parts of the humid tropics and, particularly at the busy seasons of the year (*see* p. 98 above), they have plenty of work to occupy them on the farms. They must be well nourished for this arduous work which even the strongest men can perform on average for only about 12 days in each month, so exhausting is it. They can naturally expect a high wage under these circumstances. It is also well to bear in mind that as porters are human beings they can fall sick and they can also become disaffected and unco-operative as workers do from time to time in all communities.

There are two crucial points which adversely affect the efficiency of human porterage; the amounts which a porter can carry are very limited, and, secondly, so are the distances over which he can travel. Estimates of head loads vary considerably according to circumstances, but Phillips (1954) has found that the optimum level is about 20 kg (44 lb) and that any increase in weight above this level leads to a disproportionate increase in energy costs, though he does concede that over short distances a load of 30 kg (66 lb) would not be excessive if the carrier returned unloaded (such a return journey is equivalent to a normal rest period). Estimates of a day's journey also vary considerably from 25—30 km (16–19 miles) in Rhodesia to 50 km (31 miles) in parts of China.

The overall inefficiency of porterage can be demonstrated in two ways, first by means of examining two examples. It is frequently stated that land-locked states (*i.e.* those which have no sea coast) are at a considerable disadvantage as regards foreign trade simply because of their considerable distance from the sea, and this was even more true when human porterage was the only form of transport available. Thus it is no surprise to read that before the building of the railway between Mombasa (Kenya) and Lake Victoria in 1901 the average cost of porterage between the lake and the coast was £250 per ton of goods carried, which was a very high figure indeed in the currency of those days. This made the development of trade from land-locked Uganda virtually impossible before the construction of the railway.

The second example is given by Gourou (1968, 159) and is illustrated in Fig. 23. During the early years of this century it was the custom of the French to transport imported goods destined for Fort Lamy (now N'djaména), near Lake Chad, from the port of Brazzaville, up the Congo (now the Zaïre) and Ubangi rivers as far as Fort Sibut and thence overland to Fort Crampel on a tributary of the Shari river. From Fort Crampel the goods could proceed at high water *via* the Shari to Fort Lamy. The overland journey between the Ubangi and Shari rivers was about 240 km (150 miles) and the total weight of goods to be carried

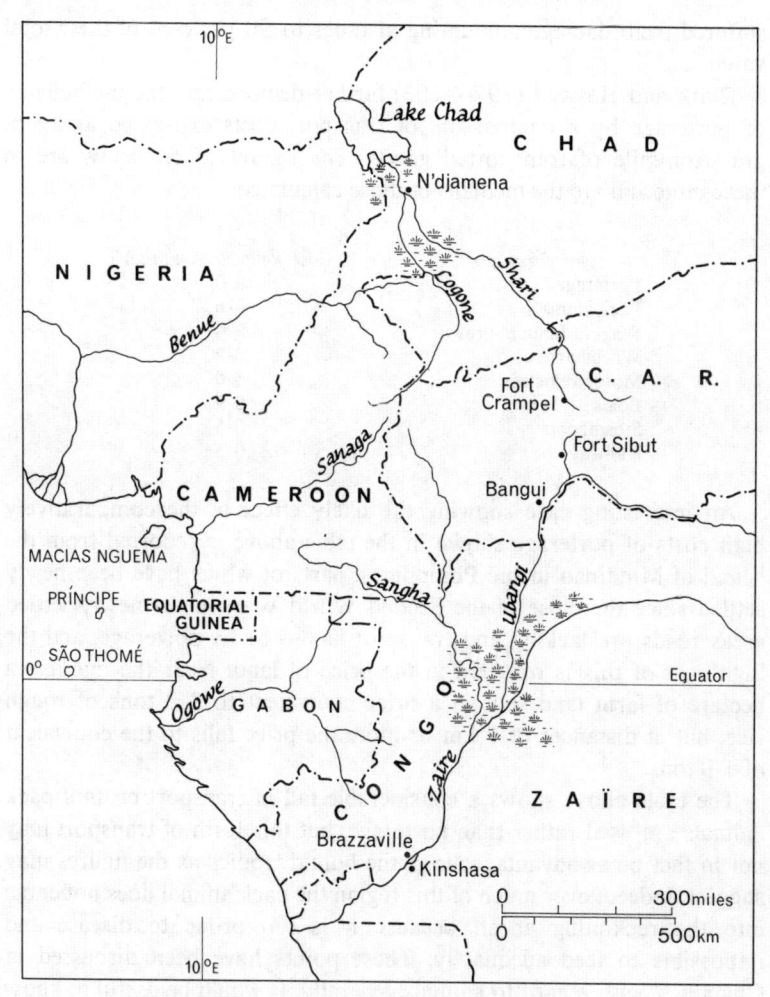

FIG. 23.—An example of transport in Africa.

each year was only about 300 tonnes, but this had to be divided up into about 9000 head-loads of about 27 kg (60 lb) each; this necessitated 100,000 man-days of porterage in all! The local population, which was sparse and which suffered from dietary deficiencies and poor health, found the consequent effort demanded from them far too great and they fled elsewhere leaving the district deserted. In all, goods travelling from Bordeaux to Fort Lamy along this route sometimes took as long as 18 months in transit; it was a very expensive journey and goods suffered from damage amounting at times to 90 per cent of their total value.

Clark and Haswell (1970, 203) further demonstrate the inefficiency of porterage by a comparison of transport costs expressed as kg of grain/ton-mile of transported goods. The figures given below are in these units and are the medians of those calculated.

Type of transport	Cost (kg grain/ton-mile)
Porterage	9.0
Pack animals	4.6
Wagons (animal-drawn)	3.4
Wheelbarrow	3.2
Motor vehicles	1.0
Boats	0.9
Steamboats	0.5
Railways	0.5

An interesting case showing the likely effect of the comparatively high costs of porterage shown in the table above is reported from the island of Mindinao in the Philippines, parts of which have been newly settled since the close of the Second World War. In the newly-settled areas roads are lacking and transport has to be by porterage, and the high cost of this is reflected in the price of land. Near the markets a hectare of farm land sells at a price equivalent to 3.3 tons of rough rice, but at distances of 6 km or more the price falls to the equivalent of 1.9 tons.

The table above shows a considerable fall in transport costs if pack animals are used rather than porterage, but this form of transport may not in fact be as advantageous in the humid tropics as the figures may suggest. Indeed, over much of this region the pack animal does not come into the reckoning at all because it is too prone to disease and impossible to feed adequately. These points have been discussed in Chapter V with regard to animals generally. It would be useful to know just how the figure given in the table was arrived at for it would seem in general terms that for the poorer communities at least the use of pack animals would show only slight advantages over porterage because of the high cost of maintaining the animals in relation to the price of labour. And pack animals, of course, are subject to limitations similar to those which affect the efficiency of human porters—the small distances per day which can be covered, the comparatively light loads, and the liability of the animals to contract disease.

On lower levels of transport costs we find wagons and wheelbarrows, though the latter are scarcely used in the humid tropics. The use of wagons is dependent upon such factors as the availability of animals (usually oxen, mules or horses) to draw the vehicles and upon the existence of roads, which must have a surface sufficiently broad and smooth to permit the passage of wheeled traffic. Variations in these factors lead to very substantial differences in price levels for this form of transport, and in general conditions in the humid tropics are not very favourable for the construction and maintenance of the necessary road surfaces. This is largely because the surfaces of dirt roads become extremely dusty in the dry season and muddy during the rains. A great deal of this loose material is lost because of the action of wheeled traffic in dispersing it and the road will therefore quickly deteriorate. There is also the danger of extensive wash-outs during heavy rain. Even bitumenised roads deteriorate comparatively quickly under tropical conditions though bitumen surfaces are not normally typical of roads designed simply to carry animal-drawn wagons. Despite these disadvantages, however, wagon transport is often typical of developing countries as it is comparatively simple in operation but it is more suited to short hauls than long; this is one important reason why it tends to be superseded by various forms of powered vehicles in more developed regions. Loads carried are limited in bulk and weight especially if road surfaces are poor while the reliability of wagon transport can be greatly reduced by hazards such as unfavourable weather, deteriorating road surfaces, or sickness among the draught animals.

One example showing the value and at the same time the limitations of this form of transport may be taken from just outside the tropics. In the early days of the Colony of New South Wales the old bullock carts, the only form of medium- and long-distance transport available, used to take about 3 months to reach the Northern Tablelands of New South Wales from the port of Newcastle, which meant that the round trip took at least 6 months. The costs of sustaining the bullock teams and the drovers on a journey of this magnitude were tremendous yet such was the demand for their services that the "bullockies" in those days were among the wealthiest members of the Australian community; there is no doubt that it was largely thanks to their efforts that it was possible to begin the opening up of these interior plateaux, but at the same time the limitations beneath which they inevitably laboured proved a severe handicap to the developing community as long as they represented the best form of transport available. In the early days the teams simply followed tracks as there were no true roads and the sheer effort involved in climbing on to the plateau, crossing in the process deep, wooded valleys holding swiftly-flowing streams alternating with steep-sided ridges, was truly tremendous, especially since the tracks rapidly became deep in mud after heavy rain. At a later date the journey became easier as the roads were surfaced and greatly improved but the bullock teams were finally dispersed only with the coming of the railways.

Water transport is almost always low-cost and in many parts of the humid tropics it plays an important part in community life. This is particularly the case in lowland regions or on the plateaux across which rivers flow smoothly over long distances as they do, for example, over much of northern Brazil, in Nigeria, in Zaïre and Burma. River transport is, however, subject to disabilities arising from the climate as we saw in Chapter I, the example of the Niger offering a case in point. This lengthy river together with its tributary the Benue which joins it at Lokoja permits river vessels to carry goods northwards from the Niger delta and inland for almost a thousand miles as far as Garua in Cameroun, but the journey is not possible during the dry season as there is not sufficient depth of water in the Benue. As the rivers begin to rise after the onset of the rains commercial canoes drawing little water are the first to make the journey northwards from the delta ports, and these are followed by larger vessels and finally by powered craft and barges when the depth of water above Lokoja permits. Near the end of the short navigation season (this lasts for only 2 months on the upper Benue) these powered craft and barges are the first to move southwards from Garua as the river levels start to fall, while the canoes are last to leave. Even on this great river there are very definite limitations for commercial forms of transport.

Three neighbouring territories in South America, Colombia, Venezuela and Brazil, help to demonstrate the varying and changing values of river transport. Colombia has always relied very heavily upon river transport, the main river and historic artery of communications being the Magdalena which is navigable for shallow-draught vessels as far as Neiva. The main terminus of river transport, however, is La Dorado upstream from which are the rapids at Honda though these can now be circumvented by making use of the railway built partly for that purpose. The River Magdalena has developed into a major means of communication in Colombia despite a whole series of drawbacks which include the rapids just mentioned, shifting sandbanks particularly in the lower river, and very marked variations in the depth of water which at times leads to flooding while at other times it means that there is insufficient water to float commercial vessels. Frequently traffic even on the lower river comes to a standstill because of very low water which may last for periods of between 6 and 8 weeks, though sometimes during such a spell a steamer can move from sandbank to sandbank as the water-level is temporarily raised by a heavy rainstorm over one or other of the upper tributaries. In the flood season, on the other hand, the current is so powerful that it may take up to 5 weeks to navigate a vessel from the mouth of the river to La Dorado, a distance of roughly 805 km (500 miles). Erosion during this season is so marked that large masses of soil, which later help to produce the troublesome sandbanks, and vegetation slide into the river even along the tributaries, and floating trees can form a menace to shipping.

Through neighbouring Venezuela flows the Orinoco, one of the great rivers of the continent, yet this waterway has never been as important a line of communication as the inferior Magdalena because for the most part it flows through little developed regions—the Guiana Highlands and the Llanos (p. 26 above). Since the 1950s, however, two United States steel companies have begun mining operations in Venezuela and they use the lower Orinoco as a means of transport. The Bethlehem Steel Corporation ships ore from Palua, on the right bank of the Caroni where it joins the main river. From the river port the ore moves downstream and across the Gulf of Paria east of Trinidad to Puerto de Hierro on the Paria Peninsula, from where it is taken by mammoth ore carriers to Sparrow's Point, Maryland (*see* Jarrett, 1974b, 186). The United States Steel Corporation ships its ore from Puerto Ordaz across the Caroni from Palua, *via* the Boca Grande en route to Morrisville, Pennsylvania, or to Mobile, Alabama.

Our third example, Brazil, is remarkable as it has the greatest of all rivers, the Amazon, though this river is not quite as long as the Nile. The Amazon begins high up in the Peruvian Andes only 115 km from the Pacific coast as a rivulet emerging from the tip of a glacier, but it rapidly increases in size. For the greater part of its length its farthest banks lie at least 100 km apart while it is no less than 333 km (207 miles) wide at its mouth where it discharges more than 200,000 m^3 of water per second into the Atlantic and floods the ocean with fresh, muddy water as far as 170 km off-shore. It is navigable for ocean-going vessels as far as Iquitos, Peru's "Atlantic port," a distance of just over 3600 km (2300 miles) while along the lowest 1610 km (1000 miles) of its course it has an average depth of 31 m (100 ft). This remarkable river, however, flows through one of the most sparsely populated and least developed regions in the world and apart from two companies, one Brazilian and one British, which operate steamers on the waterway, traffic is limited almost entirely to small boats. This situation may begin to change if the attempts now being made to open up northern Brazil (*see* p. 195 below) are successful.

These three examples very clearly demonstrate that the value of a river for transport is by no means simply a function of its physical properties but rather of the development of the regions through which it flows. As development strengthens, as in Venezuela, a previously little-used river can assume very considerable importance as a channel of communication; the fact that such a river may suffer from physical disabilities is of secondary importance if the demand for transport is sufficiently strong.

As a final observation in this section we may note that in some parts of the humid tropics lakes are greatly used for transport, the East African lakes being outstanding in this respect.

This brings us finally to the motor vehicles and the railways, both of which have a place in the table on p. 188. No one should underestimate the tremendous contribution which has been made by railways in the past to the opening up and development of the lands of the humid tropics, particularly with regard to the exploitation of mineral reserves and agricultural potential. Among the former type of development we may note the tin of the Jos Plateau in Nigeria; the copper of Shaba (formerly Katanga) in Zaïre and of the Copper Belt in Zambia; and the iron ores of Sierra Leone and Liberia. Among the latter we may include the production of groundnuts and cotton of Northern Nigeria; the cotton of Uganda; and the cocoa of Ghana (the Gold Coast in earlier times). In more recent years, however, the railway seems to have lost a great deal of its former importance as the closing down of lines in West Africa (Sierra Leone) and East Africa (Tanzania and Uganda) demonstrates though this generalisation does not hold with regard to the exploitation of bulky minerals which still need railways to move them from mine to port.

The early railways were built in the absence of any other form of effective long-distance transport with an eye to opening up particularly productive land such as the cocoa lands of southern Ghana, the groundnut–cotton savannas of northern Nigeria, or the Kenya Highlands. Sometimes, as in the case of Sierra Leone, they were built primarily for strategic reasons. The immediate effect was greatly to reduce transport costs and to make possible the growing of crops of comparatively low unit value such as sisal in areas near to the track. Crops of higher unit value such as cotton and coffee could successfully be grown farther away from the railway and transported to it from the farms. The railways undoubtedly helped tremendously in opening up the areas through which they passed for no other form of transport could at that time compete with them.

With the advent of motor transport, however, a change was inevitable. A road is a far less capital-intensive form of transport than a railway which, as we have noted earlier, must be constructed fully and all in one piece or not at all. But roads can be of varying grades from being simply broad earth tracks completely unmetalled, through those with a gravel surface, to those with macadamised and bitumenised surfaces. The critical factor is the amount of use made of the road. Earth and gravel roads are significantly cheaper to construct but they wear out quickly and are much less efficient in operation than bitumenised roads. Clark and Haswell (1970, 209) quote estimates made by Healey in India which suggest that operating costs along an earth road are approximately double those along a bitumen road. On the other hand it is not economic to pave a road unless this expensive procedure is justified by a minimum level of use; the same two writers believe that it becomes commercially worth while macadamising a road if it is used by at least 315 vehicles a day and bitumenising it for 410 vehicles over the 24 hours, while Hawkins (1958) suggests that bitumenisation in Nigeria becomes economic when a traffic density of 300 vehicles a day is reached. Pedler (1955) puts operating savings in West Africa at 25 per cent when an unmade track is upgraded to an earth road capable of carrying 15-ton lorries, and at a further 18 per cent when an earth road is given a surface of bitumen.

It is, of course, possible to economise on road construction by surfacing part only of the carrying surface, either in length or width. Surfacing, in other words, can be carried out over increasing lengths of road on the one hand as the need arises and as circumstances permit, while on the other hand it is possible as in parts of Rhodesia simply to bitumenise the two strips along the road along which the wheels of vehicles normally travel. This kind of economy is not possible in railway construction.

Roads are much more flexible in use, as well as in type, than are railways. For instance, a trunk road can be connected with villages or towns which lie off its direct course by feeders of varying grades depending upon the amount of use demanded of them, and traffic emerging from these feeder roads can proceed directly on to and then along the trunk road. And almost any sort of vehicle can use the road to suit its own particular purpose. Use is not confined to the running of trains of determinate length and constitution along predetermined routes at particular times of the day. Passengers and goods can easily and profitably move along a road in the same vehicle, while vehicles themselves can vary from a bicycle or a light motor cycle to a "juggernaut" type of heavy lorry or passenger coach. This kind of flexibility is not possible on a railway.

Grigg (1970, 86) quotes an interesting example showing the benefits which sometimes accrue from road construction when he mentions that the building of a paved road from Bangkok into north-eastern Thailand had the effect of closing down the old inefficient railway which had previously served the area. Further, the more efficient transport which was made possible by the road led to a considerable increase in the amount of maize grown in the regions it serves, and some of this is now exported. Mosher (1966) argues that agricultural activity in Borneo is confined to the coastal regions although there are extensive areas of potentially productive land in the interior. There are, however, few transport links to encourage the opening up of this land though it has been estimated that the construction of every kilometre of new road opens up 60 hectares of new land and that every £1 spent upon roads increases the annual gross agricultural production by £4.

It is perhaps not surprising under these conditions that railway construction today does not appear to have the same effect in helping to open up territory for development as it used to, valuable as railways still are for most mining ventures. It was estimated, for example, that the Bornu Extension Railway constructed in north-eastern Nigeria between 1960 and 1964 would stimulate agriculture, including the production of cash crops, just as the earlier railways had encouraged cotton and groundnut production in earlier years over much of the Nigerian savanna belt, but so far the results of the new line have been very disappointing. O'Connor (1971, 195ff) has shown that while the original East African railways, particularly the Mombasa–Kisumu (Kenya) and the Dar es Salaam–Kigoma lines, together with the branch to Mwanza (Tanganyika, now Tanzania) exercised a profound impact upon economic development, later ones have had little such effect except where they have been built to facilitate the movement of minerals. This extreme disparity is attributed largely to the influence of roads which now serve most parts of this extensive region. It is interesting to note, moreover, that in Uganda pricing policies have for some time exercised a sharp impact upon the geography of crop production. Cotton marketing, for instance, is so organised that producers are paid a uniform price wherever their farms are and transport costs do not fall upon them, while a similar arrangement operates with regard to coffee, as the Coffee Marketing Board pays transport costs. Thus there is no incentive to producers to locate their farms near a railway. Serious doubts have been voiced with regard to the new Tanzam Railway (Jarrett, 1974a, 397) which some authorities fear may become Africa's "biggest political white elephant," especially if there is a significant fall in world demand for Zambian copper.

On the other hand a bright future is confidently predicted for the proposed Trans-Gabon Railway, the first stage of which from Owendo up the middle Ogowe valley to Booué and thence to Franceville should be completed by 1980. This will make possible increased mining of manganese and uranium near Franceville and at near-by Moanda in the far south-east of the country, while an extension to Belinga in the far north-west will make possible the mining of the very large reserves of high-grade iron ore which exist there. Concurrently very large amounts of timber will be produced from the hitherto virtually untouched rain forests of the interior which include the famous *okoumé*. *Okoumé* timber which constituted the main export from Gabon for many years yields a faintly rose-coloured scented soft wood particularly well suited to the manufacture of plywood and veneers.

It is a significant point that as part of the efforts now being made to open up her vast interior and to forge better trading and social links with her neighbours, Brazil is placing the emphasis upon road construction and not upon railways. The modern age of road construction in Brazil may be said to have begun by 1960 by which time a road 2123 km long was built linking the port of Belem with the capital, Brasilia. The areas along the new road rapidly attracted settlers and an original population of about 100,000 rapidly increased to one of 2m. The further construction of 2000 km of feeder roads has helped to spark off a great expansion of agriculture, especially cattle rearing, and there may now be 5m head of cattle along this narrow strip of country while a timber-producing industry is also well established. Traffic in fact has increased in such a spectacular fashion that the original improved dirt road has now been surfaced.

The success of this project encouraged the initiation of comparable schemes and in the years 1965–76 the length of paved roads in Brazil rose from just over 26,000 km to 71,000 km while further construction is going ahead. In the north the Trans-Amazonian Highway is to link the existing network in the north-east of the country between Recife and Belem with Peru, passing right through the vast Amazon lowlands *via* Jacareacanga, Prainha, Rio Branco and Cruzeiro do Sul, while another link will run northwards from Brasilia to Santarem, Manaos, and on to Caracas in Venezuela. A further highway from Manaos will link up with Bogota in Colombia and links are also planned with the Guianas and Surinam. Nor will the extensive south-western parts of the country be overlooked as roads are to be constructed to Cuiaba and Campo Grande. In many respects the most spectacular road now open is the 500-km coastal highway linking Rio de Janeiro and Santos which cost £500,000 a km to construct, such was the difficult nature of the terrain through which the road passes.

It should not be inferred from the above that the day of the railway is over in Brazil. It is not. There are plans to construct new railways in the more densely-populated regions; the most important of the new lines will be an 830-km long railway linking Belo Horizonte in the mineral state of Minas Gerais with the iron and steel town of Volta Redonda and the industrial centre of São Paulo. This line will have a maximum gradient of 1 per cent and wide-radius curves to permit trains to run at speeds up to 130 km (81 miles) per hour, while it will include 130 km of tunnels and 70 km of viaducts and bridges. In 1979 when this line should be open the journey between Belo Horizonte and São Paulo, which now takes 40 hours, should be cut to 6 hours.

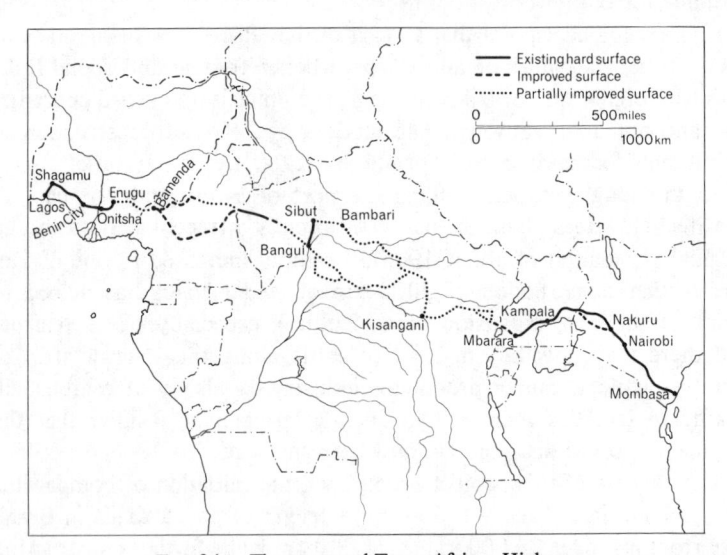

FIG. 24.—The proposed Trans-African Highway.

A programme of road construction comparable to that of Brazil is the proposed Trans-African Highway shown on Fig. 24. This will link the ports of Lagos and Mombasa though the route to be taken has not yet been finally decided; some possible alternatives are shown on Fig. 24, which also shows the vast gap which at present exists between hard-surface roads in Nigeria to the west and Kenya and Uganda to the east. The road, if and when constructed, will prove of enormous benefit to central Cameroun, the Central African Republic and northern Zaïre as it will permit intra- and inter-territorial trade on a scale hitherto unknown in these extensive interior regions, while it will also allow a much greater movement of goods destined for shipment overseas. It is also hoped that tourism will be greatly stimulated by the new road.

Developments of a similar nature are also envisaged for West Africa, particularly with regard to the proposed Atlantic Coast Highway and the Trans-Sahelian route. It is one of the legacies of colonial rule that roads and railways were planned very much on a territorial basis though this was rather less so in the extensive French West Africa than in the fragmented British territories (*see*, for example, p. 206 below). Trunk roads almost always consisted of central feeders to coastal ports (which were frequently also the administrative centres of the Colonies concerned) as the need for inter-territorial trade was hardly recognised in those days. There is thus a great need today for routes which will remedy this deficiency and the two proposed trunk roads will play an important part in doing this. Much of the Atlantic Coast Highway is in fact now well established, particularly in the east between Lagos and Abidjan, but farther west, especially between Abidjan and Bissau, much remains to be done. Much more is awaiting accomplishment along the course of the proposed Trans-Sahelian Highway between Dakar and N'djaména (formerly Fort Lamy), and particular difficulties arise because long stretches of the road will pass through sparsely populated and remote country and because it will pass through territories which are among the poorest in the world; it will in fact pass through Senegal, Mali, Upper Volta, Niger, Cameroun and Chad. Roads of this kind must be seen as long-term ventures but their future impact upon local and regional economies could be very considerable indeed.

MARKETS

As soon as transport facilities bring different communities into touch with each other, the likelihood exists that trading will develop between them, either on a local or a wider scale—or, quite possibly, on both. It is now generally appreciated that external trade forms an important part, perhaps even the basis, of economic development. Not only is the market-place a locality for the interchange of commodities but it is also one for the interchange of ideas through personal contact and it is through such means that pressure for change is generated. Trading provides a most important means of initiating and encouraging economic and social development while it also makes possible the exploitation of resources which may have limited or no value in the local community.

On a local scale, trading typically gives rise to the emergence of the day market, at least in the early stages of development, a feature which has been studied in West Africa by Hodder (1973). Hodder comes to the conclusion that for the establishment of indigenous markets in

West Africa three elements are necessary: there must be a sufficient degree of political control to guarantee law and order in the market area; there must be a sufficiently high density of population in the region served by the market; and the market itself must be located on or near a long-distance trade route. The last point reminds us that even local markets are set in a wider context than simply that of the local society. This is inevitable, for it is unlikely that neighbouring villages will have much if anything to exchange among themselves for they are likely to be producing similar crops. They will be more interested in exchanging their local products for essential goods from outside their own area which they cannot produce themselves, commodities such as clothes, household articles, various implements and seeds for the farmers to name but a few examples. The farm crops which are traded in exchange for these goods are likely to be destined for various towns to provide food for the urban dwellers or even for export.

The first and lowest stage in the market hierarchy is the periodic market which has been examined by Stine (1962) and which is of widespread occurrence throughout the humid tropics. The circumstances which give rise to the periodic market are illustrated in Fig. 25 in which we assume that there is a market demand for traders' goods over certain areas bounded by *thresholds*, while producers are able to travel with their locally-produced goods or services over areas bounded by *ranges*: it is of course more than likely that the producers concerned are in many instances also the purchasers of the commodities offered by the traders at the market centres. There is no reason why the two types of area, traders and producers, should coincide in extent though they may do so.

In Example 1, Fig. 25, we assume that the two types of area do in fact almost correspond in size, though the range lies slightly farther from the centre than does the threshold. Under these circumstances all market demands can be met if the market is held near the centre of the region concerned and its frequency will be determined by local conditions. In Example 2A we assume that the producers cannot cover the whole of the threshold-bounded area and if the market were simply held near the centre a large peripheral region would have no access to it. The only way to cope with this situation is to rotate the market, perhaps as suggested in Example 2B, so that most of the threshold area is covered. The areas not covered might well be included in adjacent market areas just as the ranges in the example shown penetrate over the threshold given on the diagram. In this case a market centre could expect to hold a market with a 3- or possibly a 6-day periodicity. Similarly in Examples 3A and 3B a 5-day periodicity could be expected.

- Market centre
- - - Threshold
— Range

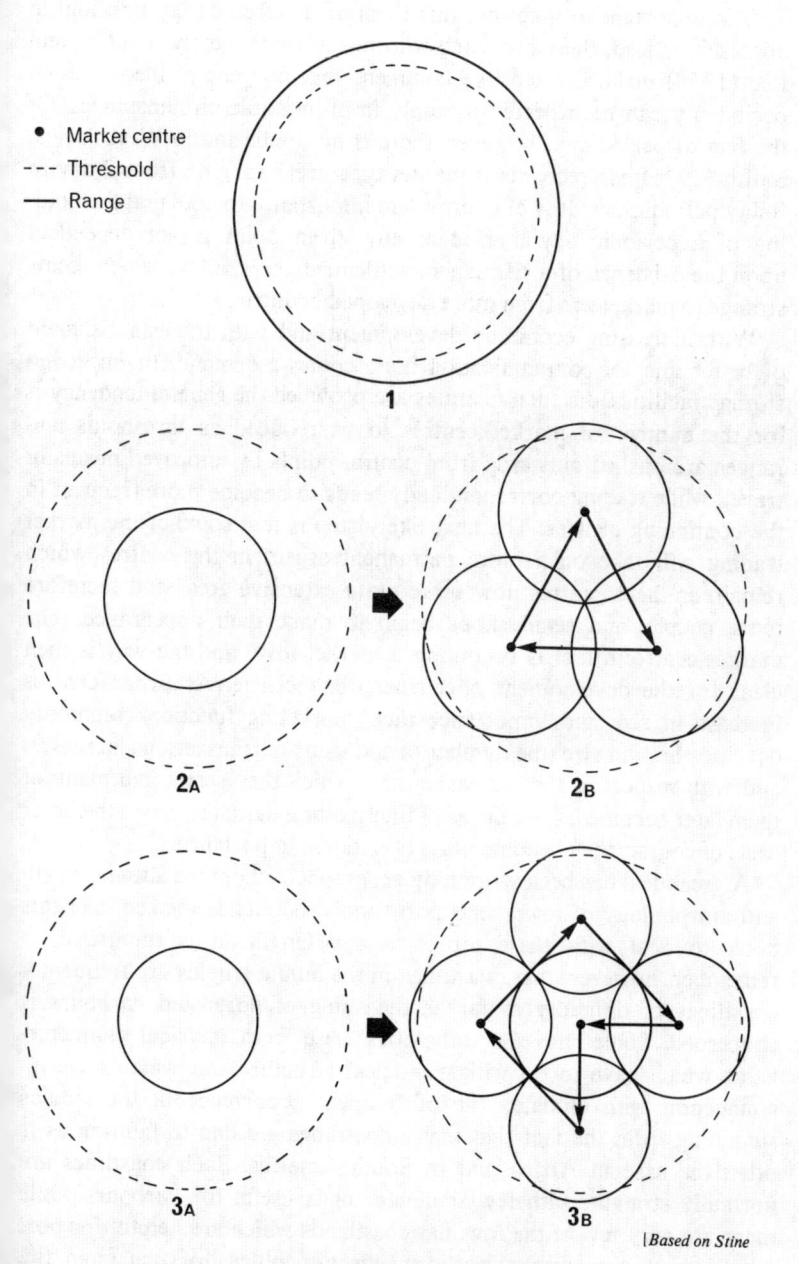

FIG. 25.—The theory of periodic markets.

Based on Stine

It is important to note that this theoretical scheme may not hold in all cases; indeed, there are some who reject it completely. Hodder and Lee (1974) make the sensible comment that no general theory about periodicity can be expected to apply in all possible circumstances. Of the fact of periodicity, however, there is no doubt and in West Africa, Smith (1971) has recognised market-type areas ranging from 2-day to 8-day periodicities. It is of interest and importance to note that the holding of a periodic day market at any given point is not dependent upon the existence of a permanent settlement there, a fact which seems strange to marketeers from more developed countries.

With increasing economic development and with the establishment of better lines of communication there comes a demand for improved storage facilities. As such facilities are provided the general tendency is for the number of market centres to be reduced as thresholds and ranges are pushed outwards from central points by improved means of travel, while trading correspondingly tends to become more frequent in the continuing centres. The next likely step is that some of the market trading will take on a more permanent aspect in the centres which remain as these centres now serve more extensive areas and therefore more people, and retail shops begin to make their appearance. The market centre in fact is becoming a market town and the way is then clear for the development of a true urban centre. As urban centres increase in size and importance their marketing function commonly develops both in size (the number of and value of transactions increases) and with respect to the increasing area which they serve, and many of them later become large centres of internal or external trade; in the latter case, of course, they become ports of national importance.

A great deal has been written by geographers about the siting, growth and morphology of towns and ports and it is outside the scope of this book to deal with these topics at any length. It is important to remember, however, that coastlines in the humid tropics are frequently coastlines of difficulty as far as the siting of ports and harbours is concerned. Sometimes the difficulties arise from physical characteristics which have to do with geological structure and which have no connection with climate, one of frequent occurrence in the tropics stemming from the fact that many coastlines are due to faulting as is often the case in Africa and in South America. Such coastlines are normally straight, with few sheltered inlets useful for harbours while there are very few of the low, flat coastlands which are helpful for port development. Sometimes, however, the difficulties do stem from the climatic environment and are due to rapid organic growth in the coastal regions. Corals, for instance, produce reefs along many tropical coasts which are therefore very hostile to shipping while coastal swamps, including mangrove swamps which are apt to develop so widely under tropical conditions, can render a coastline almost useless for port and harbour development. An outstanding example of this occurs along the West African coast of Guinea–Bissau where thick mangrove swamps give place inland to extensive coastal swamps.

Between the central Ivory Coast and the Niger delta most of the West African coast offers a classic example of what has sometimes been called a coastline of submergence with the typical off-shore bars and lagoons well in evidence. Rivers from the mainland discharge their waters and silt into the lagoons but the sandbars are broken in only one or two places. Some evidence of the strength of the various forces at work may be gauged from the fact that in 1937 a ditch 1 m wide was cut through the bar near Abidjan in order to release pent-up flood water which was reaching dangerously high levels in the lagoon. In *three days* the 1 m ditch had a mouth 100 m wide, such was the force of the released flood water, and in another 5 days this had grown to 300 m. The flow of water ceased shortly afterwards, however, and six months later the channel was completely blocked thanks to the effect of the prevailing longshore drift. It is clear that any coastline which is controlled by physical forces on this scale must present very real problems for the construction of ports and harbours and it is not surprising that provision of these essential features has been long delayed in West Africa as in other parts of the humid tropics.

TRADE

A great deal has been written about various aspects of the overseas trade of the humid tropics in earlier pages of this book. In general, most of the territories of the region have inherited a trading pattern developed in colonial times though there have been substantial modifications to this pattern in more recent years. Perhaps two main points are particularly worthy of discussion: the deterioration in and the uncertainty surrounding the terms of trade experienced by the L.D.C.s since the close of the Second World War and the attempts which these territories are making to diversify their economies and therefore their exports.

Deteriorating and Uncertain Terms of Trade

For the most part the territories of the humid tropics have had to face deteriorating terms of trade (*see* footnote p. 164 above) since the close of the Second World War and the main reasons for this were given in Chapter VII, while the special case of cocoa is illustrated in Fig. 18. We noted, as an example of this trend, that in the early 1960s a Ghanaian farmer had to market 3 tonnes of cocoa to buy a single tractor manufactured, say, in the U.S.A. or in Britain, whereas in the early 1970s the figure had risen to 10 tonnes. One reason for this deteriorating situation, which was not so serious when Crawford wrote (p. 164 above), is inflation in the industrial territories the causes of which are variously identified as fiscal, monetary, ethical and even moral. But whatever the causes there is no doubt at all about the results.

The countries which suffer from this worsening situation and with which we are now concerned are making strenuous efforts to cope with this deteriorating trade situation. They are striving in many cases to reduce their dependence upon imported food while there is something approaching a scramble to discover reserves of minerals which can be sold on world markets. Two of the smaller countries which have been particularly successful in this respect are Brunei, mentioned earlier as an important present-day exporter of mineral oil and natural gas, and Togo which now ranks as the world's sixth leading producer of phosphates. Production rose from just under 57,000 tonnes in 1961 to 1,800,000 tonnes in 1971. In 1974 the total estimated exports amounted to 2,400,000 tonnes.

There are signs at the present time, however, that increasing world consumption generally, due in its turn partly to increasing affluence and partly to a rapidly expanding population (see Chapter III above), is beginning to make itself felt in a rise in prices for primary products, and any such development must help the territories of the humid tropics. Unfortunately, however, such a progression is by no means simple, automatic or smooth; it proceeds in a very uneven manner and this can be painful and damaging. Producing states frequently attempt to benefit from such a changing pattern of supply and demand by increasing prices and the tendency is to increase by too great a sum. Demand then flags as purchasers find themselves in difficulties and there is confusion because of temporary over-production. While in the short run producer states may benefit from this financial seesawing which at times can be extremely violent, in the long run no one gains from such uncertainty and from the ill-will which inevitably accompanies it.

This continuing state of economic mayhem and the violence of price movements which accompanies it is so important a feature of contemporary economic life that we should examine it more closely. The case of mineral oil is a well-known case in point, but perhaps it is less well appreciated that movements in phosphate prices have been surprisingly comparable to those of petroleum, and in a different way the effect of this is equally damaging. Since the beginning of 1974 prices of phosphates have more than tripled, rising from about £5 a tonne to over £16 for some grades, with comparable rises in other sectors of the market and this must have damaging repercussions in territories, including many in the humid tropics, which desperately need fertilisers. Present signs, however, are that the prevailing prices are too high, that producers are experiencing selling difficulties, and that prices may have to fall, perhaps quite significantly. Even more dramatic instances include cocoa and sugar. Cocoa rose from £481 per ton in April 1973, to £640 on the 24th May. On 29th May the price had fallen to £545 before rising again and early in 1974 the price briefly exceeded £1000 per ton before sharply falling again; in July 1975, it had fallen back to £502 per ton. In the case of sugar a world shortage in 1974 caused prices on the London terminal market to rise sharply from just over £140 per long ton (itself until then a record price) in January through £250 in June to £650 in November! By the end of the year, however, the price had fallen to about £450 and it later fell to £260 and then back as far as £146. There are, of course, several points of difference between the production of crops and minerals respectively.

1. Exporters cannot fix prices of crops like cocoa and sugar on world markets as mineral exporters have been able to do. Production is much more widespread and of course tropical crop producers have to take account of competition from growers of temperate latitudes as we may note in the case of sugar.

2. Reserves of minerals are fixed and cannot be increased. This is not to say of course that all possible reserves are known at any given time, but it does mean that an upper limit is placed upon production. This is vital from the point of view of the producer country which normally has a fair idea of the length of time that it can depend upon revenue from this type of wasting asset, and such a country will endeavour to extract the maximum financial benefit from its mineral production while it can—a point examined in some detail in Chapter II. On the other hand no fixed boundaries are set to the production of a crop like sugar within certain broad environmental limits, and in the case of sugar these limits are very wide. Sugar cane, for instance, will thrive on a diversity of soils ranging from heavy clay to light sands while rainfall totals can vary between 760 and 3050 mm (30 and 120 in.) *per annum* though at the lower levels irrigation is frequently helpful. When the cane is at the strongly growing stage optimum temperatures lie near 30°C

(between 85° and 90°F) though this should gradually fall to about 15°C (about 60°F) as the time for harvest approaches. Monthly rainfall should be on the low side, about 76 mm (3 in.), for the months before and during the harvest otherwise the sugar content in the sap becomes diluted; about 130 mm (5 in.) a month is adequate for the rest of the year. Within these broad limits sugar cane can be grown and many territories within the humid tropics, especially in Africa, have significantly increased their output in recent years or are planning to do so. These include Zambia, Rhodesia, Sudan, Benin, Ivory Coast and Nigeria while outside Africa Queensland also is hoping for greatly increased production. Furthermore, the sugar grower can continue producing crops year after year on the same land as long as he keeps his soil in good condition; this situation is not comparable to that of the mineral developer who extracts his mineral and leaves a worthless mine behind him.

3. Even after planting much uncertainty surrounds the final yield of a crop. Unfavourable weather can greatly cut production—indeed, it was the catastrophic European sugar crops of 1974 which were caused by bad weather which gave rise to the wild subsequent price fluctuations which we noted earlier. Every part of the sugar cane plant is susceptible to attacks from pests, the leaves from the froghopper, the canes themselves from the small moth borer, and the roots from the greyback beetle. Various fungi and bacteria of widespread occurrence can result in diseased plants. It is not surprising under these very uncertain conditions that market estimates, and therefore market prices, fluctuate so greatly. This kind of uncertainty does not affect mineral production.

There is no doubt that because of this steady overall trend of deteriorating and uncertain terms of trade the territories of the humid tropics have for many years faced a very difficult situation; indeed, it is true to say that the financial aid given by the industrial world through various philanthropic organisations has been more than counterbalanced by this recession. On the whole, however, it may seem reasonable to argue that this difficult period may well be coming to an end though it is not as yet easy to discern a sustained upward price swing amid the short-term fluctuations. But it may in fact be on the way if the global situation is not changed by a general economic recession.

Diversification of Exports

The second trend to which we referred above which has helped (and is still helping) to modify the trading patterns of colonial times is concerned with the way in which many of the territories of the humid tropics are diversifying their economies and therefore their exports. We have previously seen that economic development in any L.D.C. is bound to be patchy and this patchiness often results in an unbalanced export pattern during what we might term the early immature stage of development. Elsewhere (Jarrett, 1974a, 587) the writer has shown how this imbalance shows itself in the pattern of exports of many African territories, and reference was made to this same point in Chapter I above. A dramatic example given there was that of the Gambia where groundnuts and groundnut products account for no less than 95 per cent by value of all exports. Other examples include Uganda, where coffee and cotton together account for 78 per cent by value of all exports, Zaïre (copper 67 per cent) and Nigeria (mineral oil and cocoa together 85 per cent). Comparable examples from other parts of the humid tropics at times show a similar imbalance, for instance in Venezuela where mineral oil and oil products account for 94 per cent of all exports, while 95 per cent of the exports of Colombia consist of coffee, mineral oil and bananas, 83 per cent of those of Trinidad of mineral oil and sugar, and 90 per cent of those of Sri Lanka of tea, rubber and coconut products.

This situation is a precarious one for the exporting country partly because of the possibility of poor harvests (in the case of agricultural products) and partly because of the ever-present danger of a fall in demand for the products concerned, possibly because of an increasing use of substitutes. A further important point is that demand elasticities may be such that price increases may result in a more or less severe fall in demand; in such cases it is necessary to impose a more than proportionate price increase in order to secure any given increase in income from any individual export, but this course of action carries with it obvious dangers. Some exports, of course, can support price increases while still retaining their buoyancy on world markets; mineral oil provides us with an outstanding case of this, and it is this phenomenon that accounts for the powerful world situation of oil producers. Even these producers, however, are slowly coming to see that their position has not altogether the strength which they once thought it had.

It is not surprising in view of all these uncertainties that the L.D.C.s are making strenuous efforts to diversify their exports; one main aim of many countries, for instance, is to increase production of processed raw materials or of manufactured goods so that they are less dependent upon crude primary products. While, as we have seen in Chapter VII, it is not an easy thing to diversify output some countries, especially the larger ones which enjoy the advantage of comparatively large domestic markets, have met with a notable degree of success. Thus, for example, 40 per cent of the exports of Brazil in 1973 were manufactured goods while primary products accounted for the other 60 per cent (8 per cent of these were minerals though coffee and sugar are easily the largest exports). In the case of India, the export of foodstuffs and raw materials now account by value for about half of all exports, manufactured goods making up the rest. Of the primary products tea alone supplies 30 per cent while iron ore is the leading mineral export. The manufactured goods show surprising variety and include both capital goods (particularly machinery for use in the textile, cement, sugar, electrical and food processing industries) and consumer goods (notably jute products, cotton textiles, leather goods, jewellery, chemicals, pharmaceuticals and oilcake).

While for many years the territories of the humid tropics have been at an increasing disadvantage as far as world trade is concerned this disadvantage has simply reflected their generally weak economic status. As they develop their own economies and strengthen their productive capacities and as their market demands intensify we can expect their trading strengths to improve, as has happened already in India and Brazil.

International Trade Patterns

We have earlier observed that during the colonial era the world channels of trade which were established for the most part linked the colonies themselves and the metropolitan countries; few, if any, trading links between the colonies themselves were developed except in such instances as occurred when a land-locked colony needed an outlet through a neighbour to the open sea. Such an instance led to the construction of the Dakar–Bamako railway which made it possible for the interior territory of French Soudan, now Mali, to share, albeit modestly, in overseas trade. When patterns of trade are as firmly established as these were they cannot be changed overnight and it is still true that inter-territorial trade within the humid tropics is still of modest proportions though attempts are now consciously being made to change this situation. A good example to illustrate this is offered by present and proposed developments in West Africa.

A Case Study: Trade in West Africa

It is not always realised what an extensive and varied region West Africa is. From west to east, from Senegal to the Cameroun Highlands, the distance is about 3200 km (2000 miles) which is roughly the straight-line distance between Dublin and Moscow; the direct north–south distance from Timbuktu to Takoradi is about 1280 km (800 miles) which is about the same as that between London and Madrid.

Physically there are wide variations between rugged mountains such as the Hombori, plateaux of varying heights such as those of Futa Jallon or Jos, river valleys such as those of the Senegal and Niger, and coastal lowlands such as those of much of the Republics of Guinea and Guinea–Bissau which are frequently fringed with mangrove swamps. Climates vary from sub-equatorial to sub-Saharan while vegetation correspondingly varies between rain forest along much of the coastal strip of the Gulf of Guinea in the south and dry steppe, sometimes known as Sahel savanna, along the northern fringes of the region. Land use shows corresponding variations from the nomadic cattle-rearing economies of the north to the root and tree crop economies of the Guinea coastlands (Fig. 16).

This physical diversity is matched by a corresponding diversity in the human geography of this extensive region. For instance, the sizes of the West African territories show great variations between, on the one hand, the areal giants such as Nigeria (923,772 km^2), Mauritania, (1,085,805 km^2) and Mali (1,204,021 km^2) and on the other hand "dwarf" or "pocket" states such as the Gambia (11,295 km^2), Guinea–Bissau (36,125 km^2) and Togo (56,600 km^2). In historical terms the territories include at one end of the time scale Liberia, which owes its beginnings to the American Colonization Society and which achieved political independence in 1847, and at the other end Guinea–Bissau which became independent only in 1974. The remainder of this vast region formed part of two great overseas empires, British and French, though there were two formerly German territories, Togo and Cameroun, which after the First World War were administered as Mandates by Britain (western Togo and western Cameroun) and France (eastern Togo and the larger part of Cameroun). The four former British colonies were the Gambia, Sierra Leone, the Gold Coast and Nigeria, while Mauritania, Senegal, French Soudan, Upper Volta and Niger together constituted French West Africa with its administrative centre at Dakar. Chad, improbably enough, formed part of French Equatorial Africa.

The economic differences between the territories are no less emphatic than the differences so far noticed. At the head of the ladder stands Nigeria, its finances swollen by oil money, which is anticipating a capital investment programme over the next 5 years of no less than £20,000m, while on the bottom rungs stand states like Upper Volta whose *total* annual budget amounts to only about £25m. The land-locked sub-Saharan states, Mali, Upper Volta, Niger and Chad are among the poorest territories in the world and it has been said (*The Times*, 30th June 1975) that the poverty gap between these territories (which some call the Fourth World) and a Third World country like Zambia is as great relatively as the gap between Zambia and one of the Western nations.

The main problem faced by these Fourth World countries are posed by their environment and their location. The critical environmental factor is climatic, for not only is the dry season long and severe and the rainy season short (*see* graph for Kayes, Fig. 3), but at times the rains fail altogether. The fairly recent Sahelian drought has had disastrous effects upon the economies of these territories; Mali, for instance, once the region's leading meat exporter, has lost about 40 per cent of its total stock, and Upper Volta, Niger and Chad about 33 per cent each. The Sahel is now a net importer of meat and it will take years for it to regain its former position as an exporter.

The location of these territories is critical because they are landlocked and the capital cities of Bamako, Ouagadougou and Niamey are each about 800 km (500 miles) from the nearest seaport, while N'djaména (formerly Fort Lamy) is farther than that; it is about 1300 km (800 miles) from Douala and rather more from Port Harcourt. The states are thus very vulnerable to external pressure as well as being weak economically though each territory has two main coastal outlets. Mali has links with the ocean through both Senegal and the Ivory Coast, Upper Volta *via* the Ivory Coast and Ghana, Niger *via* Benin and Nigeria, and Chad *via* Nigeria and Cameroun. Even so, goods travelling to or from these territories can be held up at one of the ports quite out of control of the inland state concerned, a tragic example being the holding up in Dakar of some 30,000 tonnes of grain destined for Mali during the drought as it was beyond the capacity of the port and of the Dakar–Bamako railway to handle such an amount. The capacity of this railway is only about 1500 tonnes a week.

This is not the place to enter into a detailed discussion of the economic geography of West Africa, but even from the information already given it will be clear that the development of better trade routes would be of the greatest benefit to all the territories concerned. A presumption that there is in fact a strong basis for inter-state trade is strongly suggested by the very considerable amount of smuggling which is prevalent along the borders of every country from Senegal to Nigeria. Groundnuts find their way across the border from the Gambia into Senegal; cattle cross many boundaries—indeed, it is said that about one-half of the normal meat supply of Sierra Leone illegally moves across the border from Guinea; the smuggling of diamonds from Sierra Leone into neighbouring Liberia is of very great proportions; and Nigeria is largely involved in such illicit trade through the movement of groundnuts across its northern border from Niger and through the transfer of substantial amounts of cocoa from the south-west into Benin where a better price can be obtained prior to export. Many of these illicit movements quite clearly take place because of varying pricing policies and tariff levels in different territories, particularly as between those formerly British and those formerly French, but many reflect a natural response to the varying distribution of natural resources.

One possible way to reduce this colossal amount of smuggling, indeed the only way really to eliminate it, is for the various countries to embark upon a programme designed in due course to even out these fiscal imbalances until their impact is no greater than the impact of transport costs. Governments, like individuals, must learn not to be greedy, even with the best underlying intentions, and this is particularly the case in a region like West Africa with its thousands of miles of territorial boundaries which are not controlled except at strategic points along major roads and which cannot possibly adequately be supervised. Even aerial supervision is difficult in a region of vast distances, of frequent low cloud cover and with extensive dense rain forest. It is widely recognised that there is a real need to stimulate legitimate trade and since the close of the colonial era the first steps have been taken towards encouraging inter-territorial exchange within the region by the signing of trade agreements. Some of these have been simply bilateral but the most recent, the Economic Community of West African States (ECOWAS) includes fifteen West African territories. The following table shows some of the trade agreements which have been signed in fairly recent years.

	Mano River Union	Organisation pour la Mise-en-Valeur du Fleuve Senegal	Lake Chad Basin Commission	Communauté Economique de l'Afrique de l'Ouest	Union Monétaire Ouest Africaine	ECOWAS
Cameroun			x			
Chad			x			
Benin					x	x
Gambia						x
Ghana						x
Guinea						x
Guinea–Bissau						x
Ivory Coast				x	x	x
Liberia	x					x
Mali		x		x		x
Mauritania		x		x		x
Niger			x	x	x	x
Nigeria			x			x
Senegal		x		x	x	x
Sierra Leone	x					x
Togo					x	x
Upper Volta				x	x	x

Even a cursory glance at the chart will show that for the most part the various agreements have been signed between neighbours (*e.g.* the Mano River Union) for particular purposes (*e.g.* the Lake Chad and Senegal River Organisations) or between members of the Francophone or Anglophone groups respectively. Few if any of them have made a real impact. It remains to be seen whether the new Economic Community (ECOWAS) will develop into an effective framework for economic and perhaps political co-operation or whether it will prove to be just another abortive scheme.

FINAL COMMENTS

It is sometimes forgotten that all economic production and trade is ultimately dependent upon the demands of the small-scale consumer. Even the giant producing units of the industrialised world are engaged upon the manufacture of goods which are ultimately bought by thousands of consumers of comparatively modest means. This is of course the reason why trade and the domestic economy are so intimately linked, and the point will be clear if we imagine, for instance, that the purchasing power of consumers for some reason falls. The repercussions are bound to be that fewer commodities and services are bought, and the consequent reduction in demand will inevitably affect the manufacturers (fewer goods will be produced) and also trade (there will be fewer goods to transport).

The point can be further emphasised if we examine the case of a typical farmer from, say, Sierra Leone who, as we shall almost certainly find, shares in the production of food and cash crops, probably palm kernels, cocoa and kola. Some of the food crops will not enter into commerce in any way but will help to meet the needs of the farmer and his family throughout the coming year, but some may be sold in the local market, possibly for despatch to Freetown or to some other urban centre. The palm kernels and cocoa will probably be sold through marketing boards and will be destined for export while the kola is likely to be sold in a local periodic market either for use within the territory or to be despatched along one of the well-used trade routes to the savanna zone of West Africa where it is in considerable demand, especially by desert travellers. A complicated trading network is involved in this instance, and a great deal depends upon the efficient functioning of the periodic market.

With the money received for the various goods sold the farmer, or perhaps more likely, his wife, will buy tobacco (possibly local, possibly imported), cloth and clothes imported probably from Britain or Japan, farming implements also imported probably from Britain or Germany and, in addition, household necessities many of which again are imported. The farmer's wife may travel to market in a "mammy wagon" (a passenger-cum-freight lorry) and thereby contribute to the consumption of motor vehicles, petrol, tyres and all the other essentials associated with road transport. If the wife has initiative she may well purchase goods of small bulk but comparatively high value (cigarettes, sugar or trinkets, for example) which she will take with her to sell in the other periodic markets which she visits. She will thus share in selling as well as in purchasing activities. Multiply this single example up many thousands of times in an extensive region like West Africa, which itself is only one part of the humid tropics, and you have all the ingredients for the development of a very extensive trading network which involves a very substantial volume of goods.

One difficulty is that the example just given presents a picture which in some respects is too rosy. It masks the difficulties and uncertainties which have to be faced by the farmer, and also the anxieties and the unremitting labour of himself and his family. People like these live within the grip of economic forces over which they have no control and which they cannot understand, and this situation can be a fertile breeding ground for discontent. While it is almost certainly true to argue that further development will come—Rostow estimates a period of about 60 years before under-development can give place to economic take-off and the early stages of an economically developed community—this is no consolation to those now living and who are apt understandably to become impatient with their lot. An examination of various forms of international aid lies outside the scope of this book but enough has already been said in earlier pages to demonstrate their inadequacy in view of the enormity of the problem and in the face of difficult trade conditions. "... every penny of aid that has been given has been nullified by the collapse of raw material prices and by the unbroken increase in the cost of Western manufactures" (Ward, 1964, 63). Even as this book is being completed comes the news of a further increase of 10 per cent imposed by Arab oil producers on current world oil prices and this will cost Tanzania alone an additional £5 million annually in foreign exchange; this is a tremendously heavy burden for a country already stretched to the uttermost.

At the same time, as has been emphasised more than once in this book, the future is by no means unrelievedly bleak and we have seen that there are strong grounds for optimism. Things have changed quite significantly since Ward wrote, though there is still truth in her remarks.

Present conditions, however, do not always provide a clear guide to the future. After all, it is unlikely in the extreme that an observer of the Manchester slums during the early parts of last century (and there can have been few slums anywhere as dismal and shabby as those near the Irwell and the Medlock rivers in the early years of the Industrial Revolution) would have seen many signs of hope for the future any more than would an English agricultural labourer living through the days of the post-Napoleonic depression.

It may be salutary to recall in another connection that it is not much more than a century ago that an eminent historian was confidently predicting that with a climate suitable only for "the reindeer, the elk, and musk ox" almost the whole of Canada was predestined to "perpetual sterility," while no less lacking in conviction was a report on Arizona made to the American Congress in 1858 which stated in unequivocal terms that "The region is altogether valueless. After entering it there is nothing to do but leave." The developments which have taken place in these two territories alone over the past century could not possibly have been foreseen; there is always hope for the future where the will and faith of men remain strong, and this applies to the humid tropics no less than to any other part of the world. Already many of the older (and many who are not so old) inhabitants of the humid tropics have seen very great changes and very great improvements during their lifetimes; who is to say what further changes the next generation will see?

Bibliography

Bairoch, P., *The Economic Development of the Third World since 1900*. Translation by Lady Cynthia Poston, London.
Batchelor, J. D., Journey down the Zaire River, *Geog. Mag.*, 1975, pp. 364–71.
Bauer, P. T. and Yamey, B. S., *The Economics of Under-Developed Countries*, 1965, Cambridge.
Beard, J. S., The Classification of Tropical American Vegetation Types, *Ecology*, 1955, pp. 89–100.
Beckinsale, R. P., The Nature of Tropical Rainfall, *Tropical Agriculture*, 1957, pp. 76–98.
Blanckenburg, P. von, *Rice Farming in the Abakaliki Area*, 1962, Ibadan.
Brown, E. H., The Content and Relationships of Physical Geography, *Geog. Journal*, 1975, pp. 35–48.
Brunt, D., The Reactions of the Human Body to its Physical Environment, *Quarterly Journal of the Royal Meteorological Society*, 1943, pp. 77–114.
Clark, C. and Haswell, M., *The Economics of Subsistence Agriculture*, 1970, London.
Clarke, C. G., Urbanization in the Caribbean, *Geography*, 1974a, pp. 223–32.
    Jamaica Overflows, *Geog. Mag.*, 1974b, pp. 615–19.
Cook, O. F., *Vegetation Affected by Agriculture in Central America*, 1909, Washington.
Crawford, Sir J., Problems of International Trade in Primary Products, *Progress*, 1964.
De Vries, E., *World Population Conference*, 1954, Rome.
Durand, J. D., A Long-Range View of World Population Growth, *Annals of the American Academy of Political and Social Science*, 1967, pp. 1–8.
Farmer, B. H., *Pioneer Peasant Colonization in Ceylon*, 1957, London.
FAO, *Possibilities of Increasing World Food Production*, 1963, Rome.
Firth, R., *Elements of Social Organisation*, 1946, London.
Fisher, C. A., Metropolitan Java, *Geog. Mag.*, 1972, pp. 529–35.
Fosberg, F. R., Garnier, B. J. and Kuchler, A. W., Delimitation of the Humid Tropics, *Geog. Rev.*, 1961, pp. 337–47.
Freeman, J. D., *Iban Agriculture*, 1955, H.M.S.O., London.
Fryer, D. W., World Income and Types of Economies, *Economic Geography*, 1958, pp. 283–303.
    World Economic Development, 1965, New York.
Garnier, B. J., Some Comments on Defining the Humid Tropics, *Research Notes*, Ibadan, vol: 11, pp. 9–25.

Geddes, W. R., *The Land Dyaks of Sarawak*, 1954, H.M.S.O., London.
Gourou, P., *The Tropical World*, 1968, London. Translation by S. H. Beaver and E. D. Laborde.
Griffin, K., *The Political Economy of Agrarian Change: An Essay on the Green Revolution*, 1974, London.
Grigg, D., *The Harsh Lands*, 1970, London.
Grove, A. T., *Land and Population in Katsina*, 1957, Kaduna.
Hailey, Lord, *An African Survey*, 1938, revised 1956, London.
Hance, W. A., *The Geography of Modern Africa*, 1964, New York and London.
Harris, D. R., Tropical Vegetation: an Outline and some Misconceptions, *Geography*, 1974, pp. 240–50.
Haswell, M. R., *The Changing Pattern of Economic Activity in a Gambia Village*, 1953, H.M.S.O., London.
Hawkins, E. K., *Road Transport in Nigeria*, 1958, Oxford.
Higgins, B. and Higgins, J., *Indonesia; the Crisis of the Millstones*, 1963, New York.
Hilling, D., Open Roads for Exports and Imports, *Geog. Mag.*, 1973, pp. 27–33. Developing Markets for Developing Nations, *Geog. Mag.*, 1975, pp. 636–40.
Hodder, B. W., *Economic Development in the Tropics*, 1973, London.
Hodder, B. W. and Lee, R., *Economic Geography*, 1974, London.
Humphreys, R. A., *The Evolution of Modern Latin America*, 1946, Oxford.
Huntingdon, E., *Civilization and Climate*, 1915, New Haven.
International Bank, *The Economic Development of Ceylon*, 1962, Baltimore.
Jarrett, H. R., The Strange Farmers of the Gambia, *Geog. Rev.*, 1949, pp. 649–57.
  The Present Setting of the Oil Palm Industry with Special Reference to West Africa, *Journal of Tropical Geography*, 1958, pp. 59–69.
  Africa, 1974a, London.
  A Geography of Manufacturing, 1974b, London.
Johnson, D. G., Food and Survival, *Geog. Mag.*, 1975, p. 184.
Kirk, W., Frustration of the Indian Family Plan, *Geog. Mag.*, 1975, pp. 310–13.
Lugard, Lord, *The Dual Mandate in Tropical Africa*, 1922, Edinburgh.
McPhee, A., *The Economic Revolution in British West Africa*, 1926, London.
Manshard, W., *Tropical Agriculture: A Geographical Introduction and Appraisal*, 1974, Harlow.
Miller, A. A., *Climatology*, 1971, London.
Miller, R. R., The Climate of Nigeria, *Geography*, 1952.
Mosher, A. T., *Getting Agriculture Moving*, 1966, New York.
Moss, R. P., Soils, Slopes and Land Use in a part of South-Western Nigeria, *Trans. Inst. Brit. Geographers*, 1963, pp. 143–68.
Mountjoy, A. B., *Industrialization and Under-Developed Countries*, 1966, London.
  The Form of Industrialization, in *Developing the Under-Developed Countries*, ed. A. B. Mountjoy, 1971, London.
Myrdal, G., *Economic Theory and Underdeveloped Regions*, 1957, London.
Ng, R. C. Y., Population Explosion in the North Thai Hills, *Geog. Mag.*, 1971, pp. 255–63.

Rural Change in South-East Asia, *Geography*, 1974, pp. 251–56.
Early Warning for Fertile Thailand, *Geog. Mag.*, 1975, pp. 239–42.
Nurkse, R., *Problems of Capital Formation in Under-Developed Countries*, Oxford, First published 1952.
　The Theory of Development and the Idea of Balanced Growth, in *Developing the Under-Developed Countries*, ed. A. B. Mountjoy, 1971, London.
O'Connor, A. M., *New Railway Construction and the Pattern of Economic Development in East Africa*, in *Developing the Under-Developed Countries*, ed. A. B. Mountjoy, 1971, London.
Orr, John Boyd, The Food Problem, *Scientific American*, 1950, pp. 11–15.
Patman, C. R., Probing Potential in Developing Nations, *Geog. Mag.*, 1973, pp. 641–7.
Pedler, F. J., *Economic Geography of West Africa*, 1955, London.
Phillips, J., *Agriculture and Forestry Development in the Tropics*, 1964, London.
Phillips, P. G., The Metabolic Cost of Common West African Agricultural Activities, *Journal of Tropical Medicine and Hygiene*, 1954, Vol. 57, No. 12.
Prothero, R. M., *Migrants and Malaria*, 1965, London. Nigeria Loses Count, *Geog. Mag.*, 1974, pp. 24–8.
Reddaway, W. B., *External Capital and Self-help in Developing Countries*, in *Developing the Under-Developed Countries*, ed. A. B. Mountjoy, 1971, London.
Robequain, C., *Malaya, Indonesia, Borneo and the Philippines*. Translation by E. D. Laborde, 1957, London.
Rostow, W. W., *The Take-off into Self-sustained Growth*, in *Developing the Under-Developed Countries*, ed. A. B. Mountjoy, 1971, London.
Sampedro, J. L., *Decisive Forces in World Economics*, 1967, London.
Schimper, A. F. W., *Plant-Geography upon a Physiological Basis*, 1903, Oxford.
Selby, M. J., *The Surface of the Earth*, Vol. 2, 1971, London.
Semple, E. C., *Influences of Geographic Environment*, 1911, New York.
Shannon, L. W., (ed.), *Underdeveloped Areas*, 1957, New York.
Siddle, D. J., Planning in the Third World, *Geog. Mag.*, 1975, pp. 681–4.
Simpson, E. S., Industrial Growth for an Agricultural Nation, *Geog. Mag.*, 1971, pp. 427–35.
Smith, R. H. T., *Market Periodicity and Locational Patterns in West Africa*, in *The Development of Indigenous Trade and Markets in West Africa*, ed. C. Meilassoux, 1971, London.
Stamp, L. D. and Gilmour, S. C., *Chisholm's Handbook of Commercial Geography*, 1975, London.
Steel, R. W., The Third World: Geography in Practice, *Geography*, 1974, pp. 189–207.
Stine, J. J., *Temporary Aspects of Tertiary Production Elements in Korea*, in *Urban Systems and Economic Development*, ed. F. R. Pitts, 1962, Univ. of Oregon Sch. of Business Admin., Eugene.
Taylor, Griffith, *Australia*, First published 1940, London.
Tempany, Sir H. and Grist, D. H., *An Introduction to Tropical Agriculture*, 1958, London.

Wagner, P., *The Human Use of the Earth*, 1960, London.
Ward, B., *Towards a World of Plenty*, 1964, Toronto.
Ward, B. and Dubos, R., *Only One Earth: the Care and Maintenance of a Small Planet*, Penguin, 1972, Harmondsworth.
White, H. P. and Gleave, M. B., *An Economic Geography of West Africa*, 1971, London.
White, S., The Agricultural Economy of the Hill Pagans of Dikwa Emirate, Cameroons, *The Empire Journal of Experimental Agriculture*, 1941, pp. 65–72.
Winter, E. H., *Bwamba Economy*, 1956, Makerere.
Young, A., Some Aspects of Tropical Soils, *Geography*, 1974, pp. 233–9.

Index

Abidjan, 197, 201
acha, 144
Aden, 12
Africa, 22, 27, 42, 43, 50, 56, 75, 85, 130, 159, 200
Agricultural Revolution, 61, 96
Akosombo, 57
Alabama, 191
Amazonia, 16, 18, 38, 118, 157
Amazon Lowlands, 16, 18, 20, 39, 44, 106, 195
Amazon River, 16, 44, 120, 121, 191
Andes, 18, 20, 67, 191
Angola, 55
Anopheles mosquito, 67, 68
Argentina, 160
Arizona, 212
Asia, 22, 27, 38, 47, 50, 61, 67, 71, 72, 75, 107, 145, 159
Atlantic Coast Highway, 197
Australia, 4, 20, 23, 26, 33, 63, 160

Bamako, 206, 208
Banaba, 55–6
bananas, 90, 116, 121–8, 134, 153, 205
Bandar Seri Begawan, 181
Bangkok, 194
Bangladesh, 20, 22, 41, 115
Banjul, 28, 39
Barbados, 82, 133
bauxite, 34, 35, 55
Belem, 39, 195
Belgium, 10, 44
Belinga, 195
Belo Horizonte, 196
Belterra, 121
Bemba, 89
Bengal, 44
Benin, 88, 204, 208, 209

Benue River, 37, 190
Betsileo, 128
Bissau, 197
Boca Grande, 191
Bogota, 195
Bolivia, 55
Bombay, 12, 16, 21
Booué, 195
Bordeaux, 187
Borneo, 85, 120, 194
Bornu Extension Railway, 194
Boston, 124, 125
bowal, 44
Brahmaputra River, 47
Brasilia, 195
Brazil, 3, 10, 19, 26, 27, 35, 38, 42, 43, 50, 51, 55, 56, 57, 67, 82, 85, 87, 88, 108, 116, 120–1, 131, 145, 146, 159, 162, 170, 174, 182, 190, 191, 195, 196, 206
Brazilian Plateau, 20, 56
Brazzaville, 186
Brunei, 35, 174, 181, 202
Brunei Dares Salaam, 181
Burma, 10, 19, 20, 22, 27, 145, 190
bush fallowing, 92, 93, 97, 99

caatinga, 27
caballeros, 142
Cabora Bassa, 57
Cabrais, 85, 94, 144
Calcutta, 4, 159
Cameroun Republic, 51, 144, 190, 196, 197, 206, 207, 208, 209
Campo Grande, 195
campos, 26, 195
Canada, 212
Canary Islands, 122
capital:output ratio, 173

Caracas, 7, 195
carapace latéritique, 44, 100, 144
Caribbean, 2, 20, 23, 121–8, 153, 165
Caroni River, 191
Casamance, 144
cash crop farming, 35, 91, 107–13, 116
cash economies, 90
cassava, 48, 72, 74, 89, 90, 91, 106, 133, 143
cattle, 59, 74, 128–32, 143, 144, 207, 208
Central Africa, 16, 26, 55, 63
Central African Republic, 196
Central America, 26, 29, 63, 97, 100, 102, 122, 124, 127
cerradão, 26
Ceylon, 40, 41, 66, 79, 114, 146
Chad Republic, 7, 10, 197, 207, 208, 209
Chichen Itza, 100, 103
China, 20, 38, 118, 186
climax (vegetation), 20, 23, 95, 96
coal, 32, 56, 175
cocoa, 7, 30, 51, 60, 90, 107, 111–13, 115, 116, 181, 192, 202, 203, 205, 209, 210
coconuts, 35, 109, 205
cocoyams, 91, 106
coffee, 30, 51, 90, 115, 116, 124, 192, 194, 205
Colombia, 27, 115, 170, 190, 195, 205
Congo River, 186
copper, 7, 37, 55, 192, 194, 205
Copper Belt, 55, 57, 192
Costa Rica, 73, 122, 124, 127
cotton, 7, 51, 90, 98, 113, 116, 149, 160, 192, 194, 205
crop rotation, 144–5
Cruzeiro do Sul, 195
Crystal Mountains, 57
Cuba, 46
Cuiaba, 195

Dahomey (*see* Benin)
Dakar, 12, 29, 197, 206, 207, 208
Damodar Scheme, 41
Dar es Salaam, 155, 156, 194
Deccan, 43, 56
Denmark, 174
diamonds, 55, 181, 209

Dikwa, 144
Douala, 208

East Africa, 16, 26, 29, 61, 63, 128, 156, 192
East Indies, 20, 120
Economic Community of West Africa [ECOWAS] 209, 210
Ecuador, 122, 126, 128
education, 62, 86
entrepreneur, 61, 139, 157–8
Ethiopia, 8
Europe, 60, 75
extended family, 59–60

factors of production, 135ff
fadamas, 106
fertilisers, 36, 59, 119, 141, 147, 148–50, 151, 152, 165, 184, 203
Fiji, 56
fish, 74
Fordlandia, 120, 121
Fort Crampel, 186
Fort Lamy, 186, 188, 197, 208
Fort Sibut, 186
Franceville, 195
Freetown, 12, 22, 23, 29, 210
friagems, 18
Fulani, 128
Futa Jallon, 207

Gabon, 35, 50, 195
Gambia, 7, 28, 35, 40, 90, 91, 110, 129, 130, 131, 205, 207, 208, 209
Ganges River, 47
Garua, 190
Germany, 174, 211
Ghana, 7, 20, 27, 41, 50, 51, 57, 60, 80, 107, 112, 174, 192, 208, 209
Gilbert and Ellice Islands, 55, 56
gilgai, 46
Gold Coast, 108, 112
green revolution, 148, 152
groundnuts, 7, 35, 36, 89, 90, 106, 110, 144, 155, 156, 192, 194, 205, 208, 209
Groundnut Scheme, 146, 153, 155–7, 177

INDEX

groundwater, 40
Guatemala, 32, 101, 102
Guiana Plateau, 191
Guinea–Bissau, 201, 207, 209
Guinea coast, 30
Guinea corn, 90, 144
Gulf of Guayaquil, 20
Gulf of Guinea, 37, 207
Gulf of Paria, 191
Guyana, 10, 20, 55, 195

Haiti, 10, 83
harmattan, 30, 40, 64
Hawaii, 48
herbicides, 89, 152–3, 165
Hispaniola, 126
Hombori Mts., 207
Honda, 190
Honduras, 10, 101, 124, 125, 127
Hong Kong, 133, 159
humid tropics, 10, 11, 14, 16, 23, 27, 29, 32, 34, 35, 36, 38, 39, 40, 41, 42, 43, 44, 46, 47, 48, 49, 50, 51, 52, 55, 56, 57, 58, 59, 60, 62, 63, 65, 67, 70, 72, 73, 74, 75, 78, 79, 80, 84, 85, 87, 88, 89, 96, 98, 101, 103, 105, 106, 107, 109, 110, 121, 128, 129, 131, 132, 133, 134, 135, 136, 139, 140, 142, 143, 145, 147, 148, 149, 150, 152, 153, 154, 157, 158, 159, 162, 163, 164, 174, 179, 185, 188, 192, 201, 205, 206, 212
humus, 46–7, 88, 119
Hwang Ho, 44
hydro-electricity, 41, 56–7

Iban, 89, 95
Iboland, 100
illiteracy, 9
India, 7, 10, 21, 22, 38, 41, 42, 50, 55, 56, 57, 60, 64, 72, 78, 79, 80, 82, 85, 87, 118, 119, 128, 145, 148, 149, 159, 170, 173, 174, 182, 193, 206
Indians, 61
Indochina, 50, 120
Indonesia, 55, 85, 87, 116, 119, 130, 133
Indus River, 47

Industrial Revolution, 4, 60, 161–2, 212
Inga, 57
inland delta of the Niger, 90, 106
inter-tropical convergence zone, 22–3
Iquitos, 191
iron ore, 7, 37, 55, 181, 195
Irrawaddy River, 47
irrigation, 36, 39, 41, 44, 122, 126, 133, 135, 148, 152, 203
islands of development, 7, 53
Italy, 10
Ivory Coast, 10, 19, 51, 91, 115, 201, 204, 208, 209

Jacareacanga, 195
Jamaica, 34, 35, 55, 77, 79, 124, 126
Jamshedpur, 159
Japan, 211
Java, 29, 38, 47, 85, 133–4, 144
Jos Plateau, 192, 207
jute, 115

Kanshanshi, 37
Kariba, 57
Kayes, 12, 16, 22, 23, 208
Kenya, 81, 186, 194, 196
Kenya Highlands, 192
khoei, 44
Kigoma, 194
Kingston [Jamaica], 2
Kinshasha, 57
Kisumu, 194
kola, 90, 210
Kongwa, 155, 156, 157

ladang, 87, 95, 97
La Dorado, 190, 191
Lagos, 196, 197
Lake Chad, 39, 186, 210
Lake Victoria, 186
Lake Volta, 41
Lala, 89
Land Dyaks, 94
land tenure, 140, 141–3
Laos, 22, 82
laterites, 46

Latin America, 8, 75, 79, 80, 143, 159, 184
latosols, 44
Lebanese, 61
leguminous crop, 144–5
less developed countries, 4, 6, 7, 8, 9–11, 53, 60, 75, 78, 79, 81, 98, 148, 153, 154, 157, 158, 163, 164, 165, 168, 170, 174, 178, 201, 205
Lesser Antilles, 126
Liberia, 7, 51, 55, 120, 181, 183, 192, 207, 209
Limon, 122
line squalls, 23, 30
llanos, 26, 130, 191
Lokoja, 37, 190
London, 65
Lunsar, 37
Luzon, 19, 109

Madagascar, 19, 29, 128
Magdalena River, 190–1
maize, 35, 36, 48, 72, 89, 90, 91, 98, 102, 106, 109, 133, 150, 194
malaria, 66, 67–70, 78
Malawi, 10, 185
Malaya, 29, 38, 72, 114
Malayan Peninsula, 20
Malaysia, 30, 48, 50, 55, 77, 79, 118–20, 143
Mali, 10, 90, 106, 110, 197, 206, 207, 208, 209
Manaos, 12, 16, 18, 39, 195
Mandara Mts., 144
Manding, 144
manganese, 35, 150, 195
mangrove swamps, 19, 201
Masai, 128
Matto Grosso, 121, 131, 146
Mauritania, 207, 209
Mauritius, 77, 79, 82, 174
Maya, 100–3
mechanisation, 154–7
Mekong River, 47
Menam River, 47
métayage, 134
Mexico, 20, 87, 96, 101, 102, 179
millet, 39, 72, 90, 106, 144
Minas Gerais, 51, 196

Mindinao, 154, 188
mineral oil, 32, 34, 35, 181, 202, 205, 211
milpa, 87, 103
miombo, 27
mixed farming, 132, 143, 144
Moanda, 195
Mombasa, 186, 194, 196
monsoon forests, 20, 22, 26, 27
Mora forest, 19
Morocco, 34, 149
Morogoro, 155
Mount Makiling, 19
Mozambique, 57
multi-purpose scheme, 41–2
Mwanza, 194

Nachingwea, 155
Natal [Brazil], 68
natural gas, 32, 202
N'djaména, 186, 197, 208
Neiva, 190
New Caledonia, 32
New Orleans, 124, 125
New South Wales, 189
New York, 124, 125
New Zealand, 87
Niamey, 208
nickel, 32
Niger delta, 92, 201
Nigeria, 20, 27, 37, 51, 55, 56, 70, 89, 90, 100, 107, 130, 144, 151, 156, 162, 174, 183, 190, 192, 193, 194, 196, 204, 205, 207, 208, 209
Niger Republic, 197, 208, 209
Niger River, 37, 44, 73, 106, 190, 201, 207
North America, 33, 75
Northern Territory, 20
North Vietnam, 12

Ocean Island, 55
oil palms, 88, 91, 96, 141, 153, 181
oil refining, 182, 183
Okoumé, 195
opportunity costs, 35
optimum population, 82–3
Orinoco River, 26, 191

INDEX

Ouagadougou, 208
Owendo, 194
Owen Falls, 57
Oyo, 162

Pakistan, 135, 173
palm oil, 48, 51
Palua, 191
Panama, 20, 124, 127
Papua–New Guinea, 10
Pará, 19
Paria Peninsula, 191
Penang, 118
peons, 142
pepper, 35, 89
Peru, 80, 191, 195
pesticides, 141, 148, 152–3, 165
pe-tani, 133
Philippines, 20, 30, 79, 87, 106, 154, 188
phosphates, 34, 55–6, 119, 149, 202, 203
pineapples, 48, 90, 144
plantations, 38, 44, 50, 85, 110, 113–28
Pontine Marshes, 67
population structure, 80
porterage, 185–8
Port Harcourt, 208
Portugal, 162
potatoes, 72, 74
Prainha, 195
primary succession, 20
Puerto de Hierro, 191
Puerto Ordaz, 191

Queensland, 20, 204

rainfall anomalies, 27
Rambi, 56
Recife, 195
Red River, 47, 67, 70, 135
Republic of Guinea, 20, 35, 55, 207, 209
Rhodesia, 55, 56, 57, 99, 186, 193, 204
rice, 36, 72, 74, 89, 91, 94, 95, 97, 105–7, 109, 133, 134, 144, 150, 151, 154, 181
Rio Branco, 195
Rio de Janeiro, 132, 159, 195

Rio Tapajos, 120, 121
roça, 87
rubber, 35, 38, 48, 50, 85, 90, 108, 109, 114, 115, 116, 118–20, 149, 165, 205
Russia, 174

Sahara Desert, 6, 26, 30, 67, 90
Sahel, 39, 207, 208
San Francisco, 124
San José, 124
Santarem, 195
Santos, 195
São Paulo, 3, 47, 132, 159, 196
Sarawak, 89, 94
savanna, 23, 26, 27, 43, 46, 74, 96, 101, 103
Segou, 44
selvas, 18, 19, 20, 22, 38, 157
Senegal, 7, 35, 110, 144, 197, 206, 207, 208, 209
seral communities, 20
Shaba, 55, 57, 192
shaduf, 41
Shari River, 186
shifting cultivation, 87–104, 134, 135, 136, 140, 141, 154, 166, 167
Sierra Leone, 37, 40, 55, 65, 90, 91, 100, 107, 144, 151, 162, 181, 192, 207, 208, 209, 210
Singapore, 10, 30, 118
sisal, 192
soil erosion, 29, 48–9, 105, 130, 135
sorghum, 109
South Africa, 4, 16, 33, 57, 61
South America, 16, 20, 26, 27, 38, 42, 47, 61, 63, 122, 142, 190, 200
Sparrows Point, 191
Sri Lanka, 10, 77, 83, 205
Stevenson Restriction Scheme, 120
strange farmers, 110
subsistence farming, 107, 108, 110, 134, 140
Sudan Republic, 4, 204
sugar, 36, 72, 85, 113, 114, 203, 204, 206, 211
sugar beet, 36
sugar cane, 35, 36, 48, 116, 149, 153
Sumatra, 35, 38, 48, 85, 99

sumatras, 30
Surinam, 55, 195
Sweden, 174
sweet potatoes, 90, 106
swidden cultivation, 87-8
sylviculture, 51
Syrians, 61

Tabora, 155
Taiwan, 10
Tamils, 85, 119
Tanzam Railway, 194
Tanzania, 70, 98, 155, 192, 194, 211
tea, 51, 85, 113, 149, 205
teak, 19, 51
terms of trade, 164, 201
Thailand, 10, 22, 51, 79, 88, 95, 115, 119, 120, 194
timber, 50, 51, 89, 95, 160, 181, 195
Timbuktu, 44, 73, 207
tobacco, 48, 113, 182, 210
Togo, 85, 94, 115, 144, 202, 207, 209
Tonkolili, 37
trace elements, 71, 150, 153
Trans-African Highway, 196
Trans-Amazonia Highway, 195
Trans-Gabon Railway, 194
Trans-Sahelian Highway, 197
tree savanna, 24
Trinidad, 19, 191, 205
trypanosomiasis, 66, 70

Ubangi River, 186
Uganda, 57, 115, 186, 192, 194, 196, 205

United Fruit Company, 125, 126, 127
United Kingdom, 10, 60, 77
United States [U.S.A.], 73, 97, 113, 124, 125, 127, 148, 159, 174
Upper Volta, 197, 207, 208, 209

Venezuela, 7, 10, 27, 42, 57, 80, 130, 181, 182, 183, 190, 191, 192, 205
vertisols, 46
Vietnam, 22, 162
Volta Redonda, 159, 196
Volta Scheme, 41

West Africa, 12, 26, 27, 30, 39, 40, 44, 51, 61, 64, 65, 74, 90, 98, 111-12, 114, 115, 128, 151, 153, 160, 192, 193, 197, 198, 200, 201, 206-10, 211
Western Australia, 20
West Indies, 22, 29, 122, 153
wheat, 34, 36, 113, 148, 150

yams, 72, 89, 90, 91, 106
Yangtse Kiang, 44
Yoruba, 162
Yucatan, 100

Zaïre Republic, 10, 35, 44, 55, 82, 129, 145, 190, 192, 196, 205
Zaïre River, 20, 44, 57, 186
Zambia, 7, 37, 55, 57, 59, 89, 192, 194, 204, 208
Zanzibar, 153